Peripheral Vision

EASA Series

Published in association with the European Association of Social Anthropologists (EASA)
Series Editor: James G. Carrier, Senior Research Associate, Oxford Brookes University

Social anthropology in Europe is growing, and the variety of work being done is expanding. This series is intended to present the best of the work produced by members of the EASA, both in monographs and in edited collections. The studies in this series describe societies, processes and institutions around the world and are intended for both scholarly and student readership.

1. LEARNING FIELDS
Volume 1
Educational Histories of European Social Anthropology
Edited by Dorle Dracklé, Iain R. Edgar and Thomas K. Schippers

2. LEARNING FIELDS
Volume 2
Current Policies and Practices in European Social Anthropology Education
Edited by Dorle Dracklé and Iain R. Edgar

3. GRAMMARS OF IDENTITY/ALTERITY
A Structural Approach
Edited by Gerd Baumann and Andre Gingrich

4. MULTIPLE MEDICAL REALITIES
Patients and Healers in Biomedical, Alternative and Traditional Medicine
Edited by Helle Johannessen and Imre Lázár

5. FRACTURING RESEMBLANCES
Identity and Mimetic Conflict in Melanesia and the West
Simon Harrison

6. SKILLED VISIONS
Between Apprenticeship and Standards
Edited by Cristina Grasseni

7. GOING FIRST CLASS?
New Approaches to Privileged Travel and Movement
Edited by Vered Amit

8. EXPLORING REGIMES OF DISCIPLINE
The Dynamics of Restraint
Edited by Noel Dyck

9. KNOWING HOW TO KNOW
Fieldwork and the Ethnographic Present
Edited by Narmala Halstead, Eric Hirsch and Judith Okely

10. POSTSOCIALIST EUROP E
Anthropological Perspectives from Home
Edited by László Kürti and Peter Skalník

11. ETHNOGRAPHIC PRACTICE IN THE PRESENT
Edited by Marit Melhuus, Jon P. Mitchell and Helena Wulff

12. CULTURE WARS
Context, Models and Anthropologists' Accounts
Edited by Deborah James, Evelyn Plaice and Christina Toren

13. POWER AND MAGIC IN ITALY
Thomas Hauschild

14. POLICY WORLDS
Anthropology and the Analysis of Contemporary Power
Edited by Cris Shore, Susan Wright and Davide Però

15. HEADLINES OF NATION, SUBTEXTS OF CLASS
Working Class Populism and the Return of the Repressed in Neoliberal Europe
Edited by Don Kalb and Gabor Halmai

16. ENCOUNTERS OF BODY AND SOUL IN CONTEMPORARY RELIGIOUS PRACTICES
Anthropological Reflections
Edited by Anna Fedele and Ruy Llera Blanes

17. CARING FOR THE 'HOLY LAND'
Filipina Domestic Workers in Israel
Claudia Liebelt

18. ORDINARY LIVES AND GRAND SCHEMES
An Anthropology of Everyday Religion
Edited by Samuli Schielke and Liza Debevec

19. LANDSCAPES BEYOND LAND
Routes, Aesthetics, Narratives
Edited by Arnar Árnason, Nicolas Ellison, Jo Vergunst and Andrew Whitehouse

20. CYBERIDENTITIES AT WAR
The Moluccan Conflict on the Internet
Birgit Bräuchler

21. FAMILY UPHEAVAL
Generation, Mobility and Relatedness Among Pakistani Migrants in Denmark
Mikkel Rytter

22. PERIPHERAL VISION
Politics, Technology and Surveillance
Catarina Frois

Peripheral Vision

Politics, Technology and Surveillance

Catarina Frois

berghahn
NEW YORK • OXFORD
www.berghahnbooks.com

Published in 2013 by

Berghahn Books

www.berghahnbooks.com

Library of Congress Cataloging-in-Publication Data

A C.I.P. cataloging record is available from the Library of Congress.

British Library Cataloguing in Publication Data
A catalogue record for this book is available from the British Library

Printed in the United States on acid-free paper

ISBN: 978-1-78238-023-8 (hardback)
ISBN: 978-1-78238-024-5 (institutional ebook)

This book is dedicated to my daughters, Marta and Francisca

Os olhos da memória vêem melhor que os nossos.
Nome de Guerra, Almada Negreiros
[Our memory eyes can see better than our own.]

To distract himself, he formulated a proposition. A philosophical proposition? Maybe, but tending towards 'weak thought' – exhausted thought, in fact. He even gave his proposition a title: 'The Civilization of Today and the Ceremony of Access'. What did it mean? It meant that, today, to enter any place whatsoever – an airport, a bank, a jeweler's or watchmaker's shop – you had to submit to a specific ceremony of control. Why ceremony? Because it served no concrete purpose. A thief, a hijacker, a terrorist, if they really want to enter, will find a way. The ceremony doesn't even serve to protect the people on the other side of the entrance. So whom does it serve? It serves the very person about to enter, to make him think that, once inside, he can feel safe.

The Paper Moon, Andrea Camilleri

Contents

Figures and Tables

Figures

Tables

Acknowledgements

The writing of this book was made possible by the Foundation for Science and Technology, which provided me a six-year fellowship to develop and carry out this work. I am grateful to the Centre for Research in Anthropology, where I am a Senior Researcher, and especially to its former president, Antónia Pedroso Lima. Besides the great professional enthusiasm and encouragement she has given, I am also deeply in debt to her at a personal level, for her friendship. João Pina Cabral, with whom I had the privilege to debate some of the arguments presented here, continues to be my principal anthropological guide, acting both as a mentor as a friend.

I am also grateful to Berghahn Books and to the EASA series editor, James Carrier, who accepted and followed this project right from the start, and was always helpful, patient and supportive.

There were other important people who accompanied this journey: Nils Zurawski, a careful reader of this book; Gary Marx, whose writings on surveillance were a permanent source for reflection; and Chiara Fonio and Minas Samatas, colleagues and friends who share with me a passion for our native south-Mediterranean countries and whose works continually inspired my own. At a later stage, Mark Maguire reviewed and commented on parts of this book, and Manuela Ivone Cunha gave me wise advice throughout the writing of it.

I am also very grateful to all the public officers who collaborated with me throughout the years in which this research was carried out; their availability and kindness never ceased to amaze me.

Writing a book is a lonely task and sometimes implies some collateral damages, especially regarding the family. In my case, when I think of acknowledging my husband for his constant care, I have to recognize that besides being a great companion and the best father our daughters could have, his generosity and friendship 'make my day' every day of our lives together.

Introduction

Politics, Technology and Surveillance

In August 2005 a political campaign for the mayoral elections was taking place in Lisbon, Portugal's capital. The socialist party announced as one of its most prominent slogans that it would launch 'video surveillance for Lisbon's critical areas'. The practice had been absolutely forbidden in the past, and authorization to install surveillance devices in semi-public spaces was extremely hard to obtain from the Data Protection Authority – the entity that had the final say on such matters in Portugal. However, this electoral slogan was not effectively sustained in any concrete political plan, beyond a general idea about which areas in Lisbon might be considered 'critical'. It was unanimously criticized by the whole opposition, generating heated reactions that interpreted the initiative as an attempt to turn the capital into an Orwellian *Big Brother*, considering it a threat to the privacy of its citizens, and even a return to the kind of repressive policies that dominated Portuguese life during the forty years of the Estado Novo dictatorship (1933–1974).

It was in fact the first time in three decades of democracy that anyone had explicitly brought up the subject of public 'surveillance' (*vigilância*), a term with extremely negative social and political connotations. Talking about surveillance in Portugal, especially the kind carried out by security forces, immediately revives images of the former political police (PIDE – Policia Internacional e de Defesa do Estado, which can be translated as International and State Defence Police): informers, restrictions to freedom of assembly, police brutality and other such painful memories; a form of existence which the Carnation Revolution (*Revolução dos Cravos*), led by the armed forces on 25 April 1974, had sought to overturn and replace with a democratic state. This reminiscence explains, at least partly, why the socialist candidate's pitch received such negative reactions, further enhanced by the implicit discriminatory element contained in the diffuse notions of critical areas or problematic neighbourhoods. Which areas exactly would these be, and would this decision to install surveillance cameras not be a little excessive, given the circumstances?

The socialist candidate was defeated in that election and his slogan left a bitter taste in everyone's memory. It was the first time in the history of this relatively young democracy that a project of this kind had been proposed. It contrasted markedly with the deeply rooted common notion that Portugal is traditionally a country of 'mild customs', where there is little inclination for social upheaval and no significant history of violence – something which the persistently low crime rates seemed to corroborate – in short, a country characterized by a tone of humility (Palacios Cerezales 2011; Palacios Cerezales 2003). Even the peaceful character of the democratic revolution pointed in that direction.

In this frame of mind, the mere idea of having video surveillance in public spaces was not just absurd, it was almost offensive, and even the official data seemed to support this. As a Public Security Police superintendent once protested to me almost indignantly, 'Lisbon is considered one of Europe's safest cities!'. In other countries the upgrading of security policies involving surveillance devices can reasonably be traced to concrete phenomena – such as the tragic episode, in the United Kingdom in the 1980s, of two-year-old Jamie Bulger's abduction from a shopping centre by two teenage kidnappers (caught as a result of having been captured on film by an internal CCTV system), or the reference to a 'war on terror' as an argument frequently invoked to expand the use of surveillance devices after the terrorist attacks on the United States on 11 September 2001 (Bigo and Tsoukala 2008; Feldman 2011; Kapferer 2004a; Lyon 2003; Mattelart 2010). Portugal could not be said to register any phenomena of public distress that might explain the sudden prominence of such a measure in a political campaign.

The small incident described above contains the clues for two possible interpretations of this political initiative, which will serve as the basis for my argument concerning the relation between 'politics', 'technology' and 'surveillance' as seen through the prism of a *Peripheral Vision*. The first has to do with the fact that the project to install video surveillance stood on vague and general premises, failing to set specific goals and purposes. As a matter of fact, it can only be presumed that video surveillance was intended to reinforce the city's security – usually one of the main arguments for introducing surveillance devices in public spaces. The fact is that this was never explicitly stated throughout the campaign, and neither was combating crime ever mentioned; but just as surprisingly, the same can be said of the reactions and arguments presented by the policy's opponents throughout the whole political spectrum. From Left to Right, parties were swift to declare their objection but unable to present any valid arguments to sustain it. In other words, even though the accusation of excessive securitarianism – evoking ghosts from the recent past – may have proven to be politically effective (mainly due to the strong populist potential of appealing directly to a collective conscience still fresh in many people's memories), the opposition was just as incapable of producing concrete evidence against the use of video surveillance as its proponents had been of producing evidence to support it.

There is a possible second interpretation for this controversial political episode. Its unprecedented and isolated character makes one wonder if this was not simply a case of pure political strategy, with all the rhetoric that implies. Was this just an example of empty campaign propaganda, pure meta-discourse, an intention that was never meant to be acted on? As became clear during this research, the outrage surrounding these proposals actually obscured the fact that a law had already been passed which permitted the installation of video surveillance systems in open spaces to be exclusively monitored by state police forces. This element suggests that the discussion was more complex than one might have supposed. Why were politicians acting so surprised if the law had been already approved in Parliament with votes from the majority, including the parties whose members were now raising such a fuss?

These are questions that will be discussed in this book, which aims to understand policy-making by analysing specifically how the video surveillance implementation process developed in Portugal, combining the elements of technology (the camera) and surveillance (the act of monitoring, capturing and recording personal data). In order to make this analysis, the book focuses on a period of five years, between 2005 and 2010, which roughly corresponds to the duration of the socialist government led by José Sócrates and the implementation of a specific political programme heavily centred on the idea of 'modernization', which, going far beyond mere video surveillance, was based on the crucial role played by technological innovation.

Modernization and technology were to be combined within an extensive programme known as the Technological Plan (*Plano Tecnológico*). Ultimately, the aim of this plan was to transform – one could even say revolutionize – a wide range of economic, social and political spheres of Portuguese life. As part of this plan, Sócrates' government launched the *Simplex* programme, intended as a reform and simplification of the administrative and legislative system. Basically, it sought to transform the relationship between the state and citizens by means of a thorough debureaucratization of all administrative procedures, for instance through the creation of instruments such as the citizen's card (*cartão do cidadão*), which gathered hitherto disparate personal information into a single document.

On the educational front, two initiatives within the Technological Plan should be highlighted: the Magellan Project (*Projecto Magalhães*), designed to bring new technology to the state school system and among other things to install and distribute personal computers in schools, including the free distribution of computers to students in primary education; and a project named New Opportunities (*Projecto Novas Oportunidades*), which consisted of the creation of special educational programmes for adults with the aim of encouraging and facilitating academic and professional improvement. The construction of a high-speed rail system (TGV) to improve transportation connections with other European countries was yet another project that fit this model of technologically mediated innovativeness.

In all these cases, technology played a crucial role, becoming the primary instrument that the Sócrates government planned to use to push Portugal out of a supposed 'age-old backwardness'. But these measures also became something more than a simple plan for 'progress'. Technological devices, symbolizing everything that is modern, attained a very fashionable, fetish quality. With these new computers, trains and identity cards with electronic chips, Portugal was not just attempting to improve standards of living; it was attempting to prove its modern condition (Frois 2008, 2011a).

It is important at this point to underline that this book will not be concerned with the deeper theoretical discussions concerning modernity or postmodernity. Thus, the term 'modernization' will be used here as an emic (indigenous) category; that is, as part of a terminology that refers to the sphere of shared representations and public imagination (Abélès 2010), especially in terms of the analysis of political discourse. The book will be especially concerned with this political conception of technologically based modernization. Its relevance may be explained briefly as follows: Portugal is often described (and self-represented) as a country traditionally trailing behind what are considered to be more advanced 'European' societies. This condition of backwardness finds an echo both in general public perceptions and in the reflections and expression of the cultural and intellectual elite. In this context, the motto of modernization assumes a deeper meaning than that of a mere political slogan. It symbolizes progress, development and improvement, the solution that will once and for all change Portugal's condition as a peripheral territory, devoid of any strategic or economic relevance, especially within the European Union's sphere of influence.

Such an ambition is far from being a recent phenomenon in Portugal and should not be seen as strictly exclusive to this country. Numberless examples could be presented here of similar kinds of rationale and initiative occurring all over the globe (Escobar 2012; Ferguson 2006; Gupta 1998; Holston 2008; Holston 1989; Mitchell 2002). Nevertheless, within the particular Portuguese context, it can be said to have gained renewed impetus in policy-making during the socialist governance under Prime Minister José Sócrates, demonstrated by the already mentioned proliferation of political programmes specifically and expressly conceived to accomplish such modernization. What I will strive to show is that one of the reasons for what I consider to be an *anxiety over modernization* derives from the fact that this goal seems to have eluded Portuguese imagination for so long that it has become a kind of national fixation.

Ever since Portugal became a full member of the European Union in 1986 (European Economic Community at the time), it has strived to meet European standards of development, seeing itself as needing to recover from what was perceived as time lost during the long period of dictatorship – an excessively protective authoritarian rule during which Portugal had 'turned its back to Europe'. This brings up another crucial facet of modernization within this framework: its association with a kind of inferiority complex, which accompanies a mystification of the excellence of foreign models, inversely considering backwardness as an intrinsically national feature. The coexistence of these ideas of modernization and backwardness as categories of representation has already been amply reflected upon by scholars from other disciplinary fields; I not intend to return to that debate here. Instead, I propose a close observation of one particular instance in which the thrust of modernization played a major role, and thereby a suggestion that in this obsession with modernization can be found one of the primary reasons for the consistent failure of successive governments to put Portugal on the track of sustainable and effective standards of development.

The analysis of the video surveillance implementation process in public open areas in the period from 2005 to 2010 in Portugal, and its execution through the National Video Surveillance Programme, a title that already suggests the scope of its ambition, will make obvious that the politics and policy-making that frame it, were characterized by divergences occurring on several levels. Perhaps the most obvious instance of this characterization is the atmosphere of discord experienced by the parties directly involved in the project, reflecting amongst other things the absence of any prior diagnosis as to the necessity or potential effectiveness of such systems. Thus, the aim to strengthen security for persons and goods by means of video surveillance technology – which has remained a limited practice in Portuguese cities despite its considerable presence in several other European cities – seems above all to correspond to an impulse of following foreign examples rather than actually to respond to real internal needs. The aim of this book is not so much to emphasize this project's failure, as to show that the importation of foreign models – whether in the areas of security, education, or health – is rendered inoperative not only when the different players continue to follow their own individual interests, but essentially when the project has not itself been properly conceived in terms of its applicability to the context.

I believe that this is one of the contributions this book can make both to the field of surveillance studies in general, and to the anthropological debate on matters of security, surveillance, politics and modernization. In short, this is an ethnographic study on the implementation of video surveillance, centred not on theories of social control or phenomena of discrimination or segregation, but rather on providing an accurate description of the mechanisms involved in this type of process: how institutions interact and how plans are executed, but perhaps more importantly, how the people who have responsibilities in these matters and those who are affected by them see themselves and describe their motives and their choices.

This study will tackle these already complex issues by attempting to avoid overly theorizing them or turning them into general rules that can be extended to other areas of activity or to other countries. My method consists in learning from the existing literature whilst applying an anthropological approach to address issues that I feel have not been sufficiently studied by other disciplines. The value of this approach, and its principal asset in contributing to this debate, derives both from its method of extensive fieldwork and from the adoption of a holistic perspective on the phenomenon. Therefore, this book should not be strictly understood as a work on surveillance or closed-circuit television (CCTV). It deals primarily with the importance of technology and its social and cultural impact in a particular society.

Peripheral Vision

The metaphor of 'peripheral vision' provides a particularly accurate image of the book's main themes and purposes. The term is commonly used to describe the ability to see objects and movement outside of the direct line of vision, as opposed to central vision, which occurs at the centre of the gaze and where objects are more sharply focused. The most obvious association of the term here, therefore, are the notions implied in visibility, among which I am particularly interested in the idea that the gaze implies both an observer and an object of observation and depends on copresence; that is, the establishment of a relationship between them. But the term has been adopted mainly as a play on words that has to be understood metaphorically. By combining the terms 'periphery' and 'vision', the book's title creates an equation that incorporates different aspects and finalities.

First of all, it seeks to reflect the intention to focus our observation on a specific phenomenon – the implementation of public video surveillance – in a limited time and space; that is, Portugal between 2005 and 2010. The particularity of the object is further limited by the precise time frame, corresponding to a very specific moment in this region's sociopolitical history: Prime Minister Sócrates' government, whose political programme and strategy was conceived with the strict aim of achieving Portugal's technologically mediated modernization so as to fight a perceived peripheral and backward condition. Second, this study involves analysing a device – the surveillance camera – that is used precisely as an instrument of vision, even in the sense of scrutinizing, and its implications in terms of how economic, political and social factors converge around the notions of safety and security in the urban space.

Establishing a correspondence between the term 'vision' and the scope of this work may seem a relatively straightforward matter, in that it suffices to construct this relationship around the technological device which serves as the pretext for this study. The idea of the surveillance camera as a surrogate for or even improvement on the human eye, and all the transformations it may entail for theories of visuality, are easy to recognize. But the relevance of the notion of periphery follows a more intricate route, for it bears equal relation to three distinct problems: (1) the claim of peripheral anthropology to a place next to the anthropology of the centre; (2) the peripheral condition of anthropology within the community of disciplines that traditionally problematize the uses of video surveillance systems known as surveillance studies; and (3) the specificities of a geographically, economically and politically peripheral European country.

Just as it is a widely assumed problem that anthropology has found it hard to enter certain fields of study alongside other disciplines – such as economics, political science, history or sociology – the same can be said of the anthropologies of the periphery within the sphere of their own disciplinary field. In the introduction to a work that has already become a reference for anthropology's epistemological and methodological discussion, Gupta and Ferguson state that:

> In most standard accounts of the history of anthropological theory, the canonical narrative examines the relationship between national traditions of anthropology only in the United States, Britain and France. Other national traditions are marginalized by the workings of geopolitical hegemony, experienced as a naturalized common sense of academic 'center' and 'periphery'. Anthropologists working at the 'center' learn quickly that they can ignore what is done in peripheral sites at little or no professional cost, while any peripheral anthropologist who similarly ignores the 'center' puts his or her professional competence at issue. (Gupta and Ferguson 1997b: 27)

This has indeed been a topic of debate among several scholars (Boskŏvić 2008; Gerholm and Hannerz 1982; Ribeiro and Escobar 2006). One of the main conclusions reached through this discussion is the importance (or even the absolute necessity) that peripheral scientific production acknowledge its condition by referral to the centre. The metaphor of an island archipelago and its relation to the continent suggested by Gerholm and Hannerz (1982) has been used as a motto for this self-assertion of the periphery regarding hegemonic centres. However, I have chosen to play down the struggle for a place for Portuguese anthropology within a discipline that is dominated by Anglophone hegemony, reflected primarily in that language's pervasiveness and in the greater recognition of scientific production in English (Pina Cabral 2005).

In this work, especially considering the topic at hand, it seems more relevant to address the peripheral condition of anthropology within surveillance studies. Following Gupta and Ferguson's reasoning, it can be argued that anthropologists studying surveillance-related phenomena will soon find that they are out of their element, as academic production on these themes originates mostly from sociology and political science. Moreover, any scholar, regardless of being an anthropologist or not, quickly realizes that the dominating interest has been focused on Anglo-Saxon countries. Anyone studying video surveillance, to use an example that directly concerns this work, will necessarily have to consider the British case, the foremost example of current adoption of these devices. Questioning the legitimacy of such preference, even if it implies neglecting other regional contexts where similar phenomena also occur, may prove problematic. The inverse is not true, however; so that, to omit the British case would most likely result in charges of carelessness or ineptitude, despite this being at present a fast-changing and -expanding domain (Doyle, Lippert and Lyon 2011).

What new elements can different countries or regions bring to our understanding of these phenomena? How can a local perspective add richness to the broader analyses which are bound up with dominating paradigms and therefore tend to ignore or discard other environments? These questions highlight the relevance of the concept of disciplinary periphery, both regarding the role of anthropology within surveillance studies and regarding Portugal as a relevant case study. Whilst we can safely admit that video surveillance in public areas is present in most European cities (Norris, McCahill and Wood 2004) – the British case being, for better or worse, the most striking example (Wood and Webster 2009) – we cannot fail to notice that its impact and significance differs across the different examples. That is, even though this monitoring device is becoming generalized in public spaces, there are variations in stages of implementation and modes of usage. These differences highlight dimensions that are relevant for a socio-anthropological analysis. As will be shown, the Portuguese context turns out to be particularly revealing inasmuch as it has remained unexplored territory. It provides ideal conditions for a kind of laboratory testing – to borrow an expression from Berger and Luckmann (1967) – by contrast with countries where video surveillance systems have been in operation for some time.

'Surveillance Studies'

At the origin of contemporary debates concerning surveillance practices within the social sciences are four key authors: James Rule, with his *Private Lives and Public Surveillance* (1973), Alan Westin in *Privacy and Freedom* (1967), Gary Marx with his classic *Undercover: Police Surveillance in America* (1988) and David Lyon with *The Electronic Eye: The Rise of Surveillance Society* (1994). The efforts of these authors provide foundations for the study of the role played by new technologies in collecting personal data and in surveillance practices. Their works have remained relevant and their insights may be considered visionary. This is partly due to their continued and extensive production, which has ensured their accuracy and pertinence at the beginning of the twenty-first century.

Whether because it is relatively young, or because it involves elements that are essentially mutable – namely in terms of state safety and security concerns – the field of 'surveillance studies' (Lyon 2007) is particularly prone to radical shifts. Due to the transformations in security politics and technology usage resulting from the 9/11 terrorist attacks, it has drawn increasing attention worldwide. There has been a boom in the interest of social scientists in surveillance in its various expressions, ranging from the widespread use of video surveillance monitoring systems to surveillance practices in the workplace or even commercial uses. This attention is easily confirmed by reviewing the vast academic production in recent years in connection with, for example: technology (Monahan 2006); technology's relation with democracy (Haggerty and Samatas 2010); commercial and consumer practices (Lace 2005); the notion of the panopticon (Lyon 2006); the use of identity cards (Bennett and Lyon 2008; Lyon 2009); or global surveillance and security (Zureik and Salter 2005). Furthermore, the appearance of a specific journal (*Surveillance & Society* from 2002) and of international and multidisciplinary working groups (such as Surveillance Studies Network or COST [European Cooperation in Science and Technology] Action Living in Surveillance Societies 2009–2013) can be noted. These debates were the basis of this research, and their case studies definitely enhanced the importance of studying surveillance practices Portugal.

The debate opened by the scholars in this field, and the way in which their interrogations and hypotheses have sought to explain and problematize a whole range of events, can be described as pioneering in the sense that they have clearly identified shifts in paradigm over the past decades in terms of security policies (global and internal) and their impact on everyday life. Notwithstanding, although it is undeniable that the phenomenon of surveillance and surveillance-related practices has become a matter of academic and intellectual problematization, I found it difficult to adjust the debate to a country such as Portugal, where not only were its material manifestations very slight, but also public or academic debate was by no means comparable to elsewhere. Was Portugal an exception, or was the problem simply yet to be formulated in this context?

Ideas such as *The Maximum Surveillance Society*, proposed in 1998 by the work of Clive Norris and Gary Armstrong on the dissemination of CCTV in the United Kingdom, while certainly making for an appealing title, did not seem feasible as a general theory to describe contemporary Portuguese society. The possibility that a surveillance society of Orwellian contours was anywhere imminent would not be a plausible hypothesis to anyone. Even the United Kingdom, the country with the highest number of public and private video surveillance systems, could hardly be considered something other than a democracy, with a rule of law that protects the freedom and rights of its citizens. But could it be that even so, common citizens today feel they are under constant surveillance, as some authors seem to suggest? The idea of a generalized mass surveillance society, where people are being permanently watched and monitored by a broad and unified network of interconnected surveillance mechanisms, can only be an extremist portrayal, too bleakly determinist. I found out that I was not alone in this perception, given that my own concerns were expressed in the work of Nicholas Rose's *Powers of Freedom*, where the author clearly states that he tends to 'agree with those who doubt that we are on our way to a maximum security society' (Rose 1999: 242).

These concerns made me think that some of the underlying general concepts supporting this field of study's emergence were often too unspecific or at the very least needed to be brought into context. It is clear that it is not the same thing to consider surveillance practices in countries with a recent history of dictatorship, such as Spain, Greece, Germany or Portugal, or in countries that had a completely different political history, especially long-established constitutional democracies. This difference would become even more blatant if we were to consider some former Eastern Bloc, African or Middle Eastern countries. In other words, political legacies must surely weigh on analyses done in this field. Additionally, the question of disciplinary methodologies and theoretical background, directly linked to the need for contextualized works, was also an issue to consider.

Anthropology, Politics and Policies

These ideas will perhaps help to further clarify the peripheral vision I am presenting. This metaphor stands firstly for the effort to address a geographically peripheral context, Portugal, and to make this country a valid interlocutor in the current debate on surveillance and monitoring devices of citizens, the implication here also being that other countries' contextual specificities may similarly contribute to the broader debate. Moreover, this analogy extends to the aim of bringing anthropology closer to the field of surveillance studies.

Although this study approaches problems that were originally formulated by sociologists, political scientists or criminologists, whose works are the framework for many ideas discussed in this book, anthropological methodologies seem to me to provide an adequate means to pursue a fuller understanding of the processes surrounding policy-making and institutional interaction. Thus, the contribution that this work seeks to make is connected to its capacity to distribute its attention evenly amongst the different actors involved in the phenomenon that is being analysed. This fact alone indicates the timeliness of an anthropological study of surveillance.

The idea of vision, or monitoring, by itself becomes a theme for reflection on issues of surveillance, raising multiple questions: the 'who', 'why' and 'when' of this activity; the reasons for its mediation through this technology; and finally, its role in an equation that summons notions commonly considered radically opposed, such as control and freedom. This introductory chapter will now explore a little further the idea that vision is a source of knowledge from a broader perspective – that is, not strictly linked to the monitoring activity – by considering the more dynamic dimension of its connection to policy production, as described by James Scott:

> Certain forms of knowledge and control require a narrowing of vision. The great advantage of such tunnel vision is that it brings into sharp focus certain limited aspects of an otherwise far more complex and unwieldy reality. This very

> simplification, in turn, makes the phenomenon at the center of the field of vision more legible and hence more susceptible to careful measurement and calculation. Combined with similar observations, an overall, aggregate, synoptic view of a selective reality is achieved, making possible a high degree of schematic knowledge, control, and manipulation. (Scott 1998: 11)

The author considers the role of vision, both as a metaphor and in the strict sense of a means to obtain knowledge of a given context, through a double perspective. In his work *Seeing like a State*, from which the above excerpt was taken, Scott reflects on the importance of having focused knowledge of a given situation – focused in the sense that detailed scrutiny allows one to identify specific features – in order to gain a broad or synoptic view of particular social phenomena through the aggregation and combination of these specific elements. The subtitle chosen by Scott for this book, *How Certain Schemes to Improve the Human Condition Have Failed*, could function as a starting point for reflecting on another analytical question addressed in this book: how to approach the specific phenomenon of video surveillance in Portugal without foregoing a holistic perspective.

Therefore, the anthropological trail followed in this book will essentially combine two complementary perspectives: those provided by political anthropology and by the anthropology of policy. Let us consider Jonathan Spencer's reflections on the fall and resurgence of political anthropology:

> [I]t has become brutally apparent that our collective survival depends on the ability to understand, and sometimes to anticipate, the strange world of other people's politics. . . . To achieve this, we need to pay sympathetic attention to the workings of apparently different versions of the political in places with different histories, and apparently different versions of justice and order. (Spencer 2007: 2)

Considering Spencer's observation, it could be said that the peripheral aspect I emphasize in this book also has a disciplinary dimension. It describes the role attributed to anthropology regarding the theme of surveillance, and its intention to show how anthropological methodological, interpretative and theoretical tools, applied to the study of what can be considered a 'small place', can help to shed light on the phenomenon of video surveillance, a 'large issue' in contemporary society (Eriksen 1995).

Although this study focuses on a European country and also considers data from other states in this region, it makes sense to compare it to similar case studies applying to completely different contexts, such as James Ferguson's *The Anti-politics Machine* (1994) in Africa, or Tania Murray Li's *The Will to Improve* (2007) in Indonesia. These are especially relevant for interpreting the ethnographic data and the phenomena under analysis, since they describe contexts that are similar in terms of the political discourse and policy-making inherent to the dynamics of improvement, development or modernization.

The reason for using an epistemological framework built on such works in a study dealing with Portugal is that they focus on government programmes and initiatives, pointing out these programmes' weaknesses and shortcomings, and describing their unexpected outcomes. All the same, it would still seem more logical to analyse the Portuguese case in relation to the European context – especially when studying a phenomenon such as video surveillance – and more obvious to think of Portugal as a developed country, considering indicators about education, employment, health, technology, industry, etc. However, the description and interpretation of a governmental programme – its motivations, implementation and accidents – supersedes any aprioristic conception of the field. The relevance of Ferguson's work resides in the fact that it reflects on projects which are determined, supported or financed by external powers, and in that respect it is irrelevant whether they occur in France or in Lesotho, in Europe or in Africa.

In the Portuguese case, whenever video surveillance projects are mentioned, it is almost always using 'Europe' (or one of the larger European cities) as a point of comparison. This becomes very clear when turning to the accounts of the process made by the political parties that are behind its implementation. They always mention how foreign specialists were brought in, or how certain parts of Portuguese cities can be compared to Amsterdam's red light district, or even how financing will depend on European funding.[1] The European comparison in all of these instances suggests an idea of modernization, linking Portugal to the 'developed' world.

This study is also inspired by another approach: the anthropology of policy as described by Cris Shore and Susan Wright (1997). The outlook provided by this approach is invaluable since, as they say, 'policies are inherently and unequivocally anthropological phenomena. They can be read by anthropologists in a number of ways: as cultural texts, as classificatory devices with various meanings, as narratives that serve to justify or condemn the present, or as rhetorical devices and discursive formations that function to empower some people and silence others' (Shore and Wright 1997: 6).

It will become clear that we are dealing principally with situations of resistance more than with pure political confrontation between entities or personalities, and in this respect I once again agree with Gupta and Ferguson when they state that it is 'useful to think of resistance as an experience that constructs and reconstructs the identity of subjects. As a form of experience, resistance's effects on the identity of subjects may be profoundly transformative. But it may equally result in reconfirming or strengthening existing identities, ironically contributing to maintaining the status quo' (Gupta and Ferguson 1997a: 19).

Closed-circuit television practices have been mainly discussed for their discriminatory potential in the hands of those who operate and monitor the images – the state police or private security companies. However, this study aims to see beyond such an accusation, which frequently results in partial and biased analyses. It is concerned not so much with video surveillance as a theme in itself, but with exploring the other problems it raises. Thus, video surveillance emerges as the binding element for a number of topics

– civil rights and freedoms; the public and private spheres; Portugal and the European context. Above all, it becomes a pretext to reflect on the relations between politics, governance and strategies of power. Not only was this research carried out in different geographical areas, as it involved interlocutors from different spheres of power, gathered around the common concern to install public video surveillance, but above all, it is what Gregory Feldman calls nonlocal ethnography that 'shifts the primary object of study from location-specific practices to discourses that enable, organize and effectively integrate so many disparate policy practices beyond the locality' (Feldman 2011: 33).

One may ask how surveillance practices in a small country, with no economic or political power, can offer a relevant contribution to the broader debate. In fact, Portugal affords an ideal setting to witness the process of CCTV implementation from diverging perspectives. In this sense, this book seeks to provide a perspective on a peripheral country that offers ideal conditions for a well-grounded case study focusing, for instance, on how small security constellations are formed within modern cities. This gives us the chance to witness an ongoing phenomenon while comparing it with existing data from other countries where similar security devices have already been in use for some time. Furthermore, as will become abundantly clear, this study shows that, like in so many other political matters, we are dealing with issues which, far from gathering consensus or mobilizing shared interests, present us with disagreeing forces involved in diverse forms of power struggle.

It should also be stressed that despite the focus on technologically sophisticated and relatively recent forms of surveillance, such as video surveillance systems, this book makes no claim to originality or innovation in terms of bringing the topic of security to the forefront of academic discussion (compare Balzacq and Carrera 2006; Burgess 2010; Eriksen, Bal and Salemink 2010; Loader and Walker 2007; and Zedner 2009). I will be particularly cautious in this respect, refraining from making definite assessments as to whether we are living in a more unsafe society or whether the levels of control and surveillance are greater than two, three or four decades ago. This abstention becomes particularly justified in the Portuguese context: in a country with a history of dictatorship, where repression and intense political control dominated social life for most of the twentieth century, some of the problems currently posed regarding the use of public video surveillance systems are not new at all, finding parallels in other historical and political periods. Thus, I will not base the argument of this book on the premise that the emergence of video surveillance systems in Portugal corresponds to a recent concern with issues of security or is evidence that the country is living a period of increased insecurity. I will also seek to avoid the opposite extreme of seeing in video surveillance systems an alleged return to past practices and behaviours. To focus on the precise juncture at which events are occurring, trying to describe their development and identify underlying motives, is in my opinion an important part of the effort to gain better understanding of the scope and significance of security devices such as those considered here, regardless of the country or region that is being studied.

Notes on Methodology and Ethics

The argument developed in this book is based on ethnographic research (Lima and Sarró 2006; Melhuus, Mitchell and Wulff 2010; Pina Cabral 2006; Pina Cabral 1991) that was conducted consistently between 2007 and 2010, including interviews with officers from the various Portuguese state security forces, members of the Portuguese parliament, the Ministry of Internal Affairs, city mayors' councils, and the Portuguese Data Protection Authority. In addition to the information gathered from these sources, I perused a large amount of material, including rulings and legislation, parliamentary records, and reports and decisions on video surveillance made between 2005 and 2010.

In my perspective, identifying the multiplicity of agents and understanding the motivations involved in the operation of surveillance, identification and monitoring systems in Portugal are crucial to help us make sense of processes that have increasingly occupied a place within the interstices of our social fabric (Frois 2011a, 2011b). The voices of other interlocutors in this book were collected through research of the online edition of the Portuguese daily newspaper *Público*. This type of personal, remote and undirected information, while presenting some limitations, is a unique source of otherwise inaccessible material (Pina Cabral 2002).

This being the first study conducted on video surveillance in open spaces in Portugal, during fieldwork I was confronted with my own preconceptions regarding the subject and even my interlocutors. I started out under the misconception that it would be difficult to gain access to political agents involved in these matters, expecting either to be dismissed with the excuse that they were too busy, or be given the same rhetorical discourse usually handed out to the media. Similarly, I was certain that my request to study processes involving video surveillance at the Data Protection Authority would be denied and, lastly, that security forces would shut me out from their corporate cocoons. Somewhat paradoxically, I presumed that I would find few ambiguities and clearly defined roles: Portuguese authorities would certainly be enthusiastic supporters of video surveillance in public areas, given their notorious craving to be considered modern and their desire to 'keep up' with other European countries in the field of technological development and competitiveness (Frois 2008).

What I found challenged these preconceptions. Every single one of my interlocutors was extraordinarily generous and open to discussion. This openness, in fact, turned out to pose an ethical challenge. Discovering (once again, contrary to what I initially supposed) that this topic did not meet with consensus among the parties, implied that my management of the information, my dialogue with the different entities, and even my approach to and dissemination of the results would demand a greater degree of caution. My interlocutors belonged to distinct groups. At the same time that each followed a particular set of internal rules, goals, methods and policies, they were closely connected by the matter at hand, making them strangely interdependent.

As will be shown, the process leading up to the implementation of a video surveillance project for any particular public space moves gradually through a series of institutions that contribute to the proposal's completion, including the final stages of approval or rejection. Despite the distinct responsibilities and specific powers of each entity, ultimately they all bear the same weight, the only difference being that municipalities and police forces answer directly to the Ministry of Internal Affairs while the Data Protection Authority, due to its independent nature, is free of this restriction.

Following the process from its inception, it soon became clear that video surveillance in Portugal is principally a pretext for political dispute or institutional performance, rather than an area for effective and purposeful action. This is where the 'side effects' mentioned by Ferguson show that we are dealing with situations where power (in the sense of authority and domination) is ultimately the issue at stake. This is not the type of power concentrated or commanded by government, allegedly the maximum authority within a state, but one that is scattered and that emerges in the form of small pockets of resistance, creating sufficient interference to gradually weaken what may have initially been a strong resolve to enforce measures. Moreover, as James Scott's distinction between public transcripts and hidden transcripts helps to make clear, this kind of negotiation constitutes an end in itself. As the author explains, 'a comparison of the hidden transcript of the weak with that of the powerful and of both hidden transcripts to the public transcript of power relations, offers a substantially new way of understanding resistance to domination' (Scott 1990: xii).

Within the anthropology of policy, a tendency to emphasize the specific features of fieldwork dealing with political institutions or, taken more broadly, with any kind of elite or group of specialists is frequently found (Shore and Nugent 2002). The type of features commonly pointed out include, for instance, the degree of difficulty of access to a given terrain, or a characterization of the interlocutors' relations of power, both in terms of their status within a given institution or in their relationship with the anthropologist (a member of another elite group). In other words, it is implied that to a certain extent, the terms in which research in these contexts is conducted are necessarily conditioned more by the participants' intentions and receptiveness than by the decision made by the anthropologist in his/her choice to explore that territory and talk to those people. And yet, we must keep in mind Cris Shore's comment that '[i]f anthropologists constitute an intellectual elite, . . . they represent only one kind of professional elite among many, and their status and influence is small beer compared to that of the more "strategic occupations" (or professional gatekeepers) such as lawyers, doctors, politicians, senior civil servants and company executives' (Shore 2002: 2).

Relations of power are certainly involved in any fieldwork done with elite groups, just as the need to obtain special authorizations and endure long waits usually are. It is also true that feelings of frustration may arise from the difficulty, due to busy schedules, in accessing certain persons, whose reactions range from: (1) revealing a lack of interest in the work of the anthropologist, being accordingly tardy or unwilling to facilitate contact; (2) misconstruing the anthropologist's interest and his/her role as an informed actor, thus being concerned with presenting a good impression both of him/herself and of the institution; (3) correctly understanding that anthropologists, just like any other kind of researcher or scholar, are entitled to their fair amount of attention. Other reasons might easily be added to the alleged specificity of fieldwork done in institutional terrains that are geographically scattered, and where it is hard to generate ties of familiarity and some measure of continuity, making each successive contact seem like the first.

During this research it became clear that successful fieldwork does not entirely depend on the subjects' status or wielding of power, but also on the patience and skill of the anthropologist in the negotiations entailed (Shore, Wright and Però 2011). Hence, distinctions between 'terrains' should not be based on their degree of difficulty. What is specific to studies that deal with any type of elite group is the degree to which subjects in these contexts are aware of those features throughout interaction with the anthropologist (Lima 2003). But there is a difference between distinguishing the specific features of a certain field, finding ways in which to deal with them during fieldwork, and making a judgement on the comparative difficulty of one type of fieldwork over another. Apart from the methodological nuances, it is obvious that anthropological work done with elite groups has the advantage of providing a 'useful focus for addressing a range of core anthropological and sociological concerns including language and power; leadership and authority; status and hierarchy; ideology and consciousness; social identities and boundary-maintenance; power relations, social structure and social change' (Shore 2002: 9).

The methodological insights provided by Susan Wright (2011) and Shore, Wright and Però (2011) were also valuable when this study focused on yet another specificity of the anthropology of policy: the fact that sometimes there is no 'field', at least in the way it is commonly understood by ethnographers. Additionally, this conception of indefiniteness applied to the field extended to one of its more relevant elements, namely its locale. This is especially relevant when the anthropologist is forced to recognize that the material containing relevant data for understanding a given phenomenon may be as diverse and scattered as the interlocutors involved. Besides a methodological update, this implies a special concern with the ethics involved in dealing with interlocutors who may sometimes simultaneously be competitors or opponents.

The reason for expounding these methodological impressions stems from the need to clarify some of my own misconceptions regarding the difficulties surrounding this type of field research. I was preparing myself for a tough, inaccessible terrain, where I would be regarded as an unwelcome presence and find little receptiveness from potential interlocutors. However, the institutions I approached (in their majority public ones) presented themselves in a different light, perhaps one more in accordance with an attitude of political correctness; that is, they granted me willingly the time for an interview that was above all intended to convey the institution's policy of transparency. In a sense, the situation was similar to Marilyn Strathern's description in *Audit Cultures* (2000) of contexts where everything must remain confirmable, accessible, accounted for and open to question, but where in fact this appearance of transparency and accessibility becomes itself a form of control over the information that is produced.

In the Portuguese case, mostly due to the campaign of technological improvement, much information was being made digitally available, not only to social scientists, but to all citizens. This effort was supposed to manifest both a belief in modernization and a political investment in democracy and transparency, even if it seldom went beyond intentions. The ease of access to information and persons became extremely perceptible during this stage. To give just one example, I discovered that all the decisions and rulings passed by the Data Protection Authority were available online on this institution's website, contradicting my supposition that they would be privileged information. Furthermore, all the bureaucratic and procedural measures of video surveillance, including all documents from other entities involved, were available for consultation at the Data Protection Authority's main offices.

During fieldwork, unsuspected problems were in fact encountered, but ironically, they derived not from any sense of restriction, but rather precisely from the ease of access to public officers holding prominent institutional posts. It was not so much the potentially confidential nature of the information I was being given access to that made me feel ill at ease. What constituted a problem was the fact that the officials I contacted expressed their personal opinions all too freely, being willing to contradict – and even criticize – the positions and actions of other entities involved in these projects. Very soon I was introduced into a world where the initial tone of political correctness quickly turned into one of conflict, showing that these projects were far from being truly coordinated affairs, even internally within each institution.[2]

The kind of criticisms and divergence to which I am referring were not just meant as general comments on the system's flaws, but were often directed at specific persons and institutions. While I was surprised by the lack of prudence and reserve all this evinced, the freedom with which personal opinions were conveyed to me by different parties also had the effect of making me aware that I had to be particularly cautious so as not to incur any misunderstandings or breaches of confidence (Caplan 2003; Macdonald 2010; Meskell and Pels 2005; Mitchell 2010; Mosse 2011).

Indeed, even after a few months, I was still conscious of the need to tread carefully, since I understood that it was not my role to confront interlocutors with the evident lack of communication that I was given to see through my knowledge of the different views. In the end, what I came to realize, even before understanding what the Portuguese video surveillance endeavour actually implied, was that this was not in the least a peaceful subject, since the principal intervening parties were involved in veiled confrontations with each other, and overall they were so concerned with defending their own institutional positions that they were willing to stall the advance of the whole initiative.

Structure of the Book

The anthropological approach adopted in this study traces the vicissitudes of a political project marked by inconsistencies, omissions and contradictions. I consider that the conflicting relationship between different social agents responsible for determining the success or failure of public video surveillance projects, revealing the intricacies of this type of initiative, renders the Portuguese example relevant for a reflection on the specific subject of public video surveillance, and at the same time allows us to question the alleged hegemonic model of security at a global scale.

Chapter 1 gives a portrayal of contemporary Portugal by means of an account of its recent history spanning most of the twentieth century. This initial narrative highlights the most significant political events and turning points, ranging from the Estado Novo's dictatorship regime to what are identified in the chapter as the successive stages of the country's political, economic and social stages of modernization. By bringing the recent history of this country into context for the reader, this historical analysis provides the necessary clues for a deeper understanding of the social and cultural role played by the political discourse inspired by the themes of development, progress and modernity.

Chapter 2 plunges straight into the main subject matter of this book – the use of video surveillance in public areas in Portugal in the period between 2005 and 2010 – providing an in-depth presentation of the National Video Surveillance Programme and its implementation in Portuguese cities. This analysis anticipates that such a project, far from being as unanimous and straightforward as citizens were led to believe, was actually riddled with multiple of obstacles. As Chapter 3 makes abundantly clear, these problems arise from various sources, but they can be primarily traced to the flawed relationships among the different institutions and powers involved in the process: between the Public Security Police and the National Republican Guard, the Ministry of Internal Affairs or the Data Protection Authority.

As the study progresses, two facts become increasingly clear. The first is that importing a foreign model, without first carrying out (a) a preliminary analysis of their actual need and (b) an assessment of the existing structures and means, renders the whole process inoperable from the start. The second is that however modern and sophisticated a system might be, this does not by itself solve an intrinsic cultural and mental inability to change and adapt to it, namely that of institutional cooperation.

Taking into account the impact of the use of public surveillance systems in Portugal, considering its historical legacy, Chapter 4 makes use of material collected from media sources to try to understand citizens' perceptions of these issues, specifically as they equate notions of freedom, privacy, rights to image and free circulation. Permanently intertwined with this reflection and analysis is the discussion of the meanings attached to notions of *security* and *insecurity*, both in the public and in the private sphere, as well as the study of how political and economic mechanisms react to the strictly subjective phenomenon of fear and security, which is shown in Chapter 5.

The Conclusion summarizes some of the issues present throughout the whole work, to show that the effort towards modernization is constantly met by a cultural resistance to change that is ultimately strengthened by an intellectual discourse bent on identifying those same traits of backwardness, actually working as a kind of self-fulfilling prophecy. Thus, by suggesting that the lack of communication between political decision makers and common citizens essentially results from the confusion between objective data and subjective perceptions, this book ultimately seeks to reveal the idiosyncrasies and fragilities of Portugal's democratic system.

Notes

1 *Diário de Noticias*, 20 December 2007. This article refers to the project for public video surveillance installation in Lisbon's downtown, which, as will be examined in Chapter 2, was characterized by successive rejections by the Data Protection Authority of the parish president's continuous, but isolated and improvised proposals.

2 I find that Peter Pels' description of anthropological ethics as a technology of self is well suited to this context, particularly his claim that: 'To argue that to guard against the negative effects of the depolicitization of research by ethics and method, anthropologists may have to consider agonistic confessions, the historicizing and politicization of methodology, and the possibility of an emergent ethics, one that is no longer tied to stable communities but arises from contingent negotiation. The consciousness of political engagement, however, also brings forward the trickster's dilemma: owing public allegiance to both research sponsors and research subjects, anthropologists can no longer desire to show either of them a "true" face' (Pels 2000: 137).

Chapter 1
From Dictatorship to Democracy

Portugal, Europe's westernmost region, is situated in the Iberian Peninsula, wedged between the Atlantic Ocean and the Spanish border. Its territory, which has remained virtually unaltered for over eight centuries, makes Portugal and its population of approximately ten million Europe's oldest nation. During the fifteenth and sixteenth centuries, due to its pivotal role in the maritime expeditions that opened the routes to Brazil and India, it established a vast colonial empire which would only definitely disintegrate in the twentieth century. The gradual economic and political weakening of the monarchic regime that ruled Portugal for most of its long national history led to its replacement in 1910 by a republican state that would shortly after give way to a military dictatorship in 1928. This in turn was substituted in 1933 by an authoritarian regime that would last for most of the twentieth century, and would only be overthrown in April of 1974, in the so-called Carnation Revolution, as economic and political isolation rendered the leader António Oliveira Salazar's position increasingly vulnerable in the European context.

The military coup that paved the way for the young democratic system that has governed Portugal ever since was understood by many as finally bringing closer the age-old promise of modernity for a country marked by an eternal stigma of underdevelopment. This condition, often associated with a kind of national destiny, is best symbolized by the term *fado* (from the Latin *fatum* meaning destiny), the name of a distinctly Portuguese musical expression, akin to the Brazilian *samba* or the *morna* of Cape Verde. It is the music of unrequited love, tragedy and longing.

What can fado, or destiny, have to do with politics, technology and surveillance? Basically, this relationship serves to illustrate one of the major arguments underlying this book, which can briefly be formulated thus: in Portugal, terms such as 'modernization', 'progress' and 'development' are continually invoked by a wide range of social actors, representing the right path and ultimate goal of all political and social change, but on the other hand conceal the actual truth that, to use Latour's expression, we have never been modern. The result is that the demand for modernization is accompanied by the parallel reification of 'backwardness'. Alluding to Portugal's peripheral condition, to its distance from the rest of Europe and so forth, is part of common everyday discourse, and the country is typically portrayed as a kind of European backwater, forever lagging behind more advanced states. This fate is built as much on the renewal of belief in the potential of the Portuguese people as on their persisting failure to achieve it.

As a member of the European Union, Portugal must follow this community's directives and enforce its legislation. On another level, there is an eagerness to keep up with European trends. To some extent these represent a departure from Portugal's not so distant history of authoritarian rule, whose influence still finds many reflections in this Southern European, Catholic and conservative country, determining its specific social, cultural and economic dynamics. Perhaps the image of someone whose peculiarities render them vaguely maladjusted but nevertheless strives to fit in, may help to illustrate the point I am trying to make. On the one hand, there is an element of European political and economic hegemony, and on the other hand, an element of local specificity. The relation between these two poles is, in my opinion, one of the features that make this case an interesting one in the context of contemporary surveillance studies. The fact that a peripheral country such as Portugal is still taking its first steps in the use of video surveillance in public areas constituted a great opportunity to witness the above-mentioned phenomenon, gaining first-hand access to the process of implementation of video surveillance devices and the chance to reflect on their future maintenance.

Portuguese history is an essential aspect of the arguments presented in this book, especially considering the proposition that video surveillance implementation must be understood as part of a broader political project of modernization intended to fight this country's underdevelopment. This study aims to show the extent to which the opposition between notions of progress and backwardness is, as in all dichotomic relationships, based on a substantiation of each of its parts, or in other words, that the recognition of the need for development implies the acknowledgement of an existing backward condition. Additionally, this idea of modernity or development is established by comparison to an external model, namely that of other European countries, to which those qualities of progress and sophistication are attributed, even though many of the international indicators for the last three decades (in such important fields as health, education and security) seem to point to Portugal's fast and sustained evolution. In any case, this fact that does not seem to be accompanied by a change in the common discourse, both internally and internationally, which continues to represent Portugal as a country that is still at the 'tail of Europe'.

The main goal of this chapter is to give an account of the most significant transformations that have occurred in Portugal roughly since the mid twentieth century, showing how the indelible mark of an authoritarian legacy of dictatorship continues to produce effects on Portuguese life up until today, particularly on two major spheres: first at the cultural level of mentalities, and second at the political level in terms of political-party dynamics and institutional relations. Fully understanding the process of video surveillance implementation in Portugal between 2005 and 2010 requires paying close attention to this country's recent history, particularly to the period extending from the mid-1970s to the present. Besides helping to confirm the recurrence of the modernization discourse already mentioned, this effort helps to highlight yet another significant aspect that is directly linked to the legacy of nearly four decades of political dictatorship in Portugal.

Although it may seem all but controversial to acknowledge the relevance of historic events towards understanding current developments, the interpretation of how those events actually shape the present situation is a far more intricate matter. Thus, the aspiration to provide a clear reading of a country's recent history, and subsequently present it in the form of a theory that clarifies the present context, must start with recognizing not just that such an aim is ambitious, but also that however objective or circumscribed its conclusions strive to be, they can never attain a character of finality.

More than searching for a thematic correspondence, this analysis is interested above all in revealing the existence of a whole underlying mindset. In other words, it will not be necessarily concerned with establishing a direct analogy between a regime that relied strongly on surveillance practices and the apparent direction taken by recent political measures, with the result that this would probably limit the inquiry to determining the extent to which the latter constitutes a continuation or a break with the former, ultimately leading to overly rigid categorizations. Instead, this perspective will rely on the premise of a more flexible relationship of action and reaction, in which the past is continually called to bear on present events but nevertheless remains a strictly residual presence in contemporary democratic practice (Freire 2009).

In fact, throughout this investigation, both bibliographic research and interaction with interlocutors revealed that the Portuguese democratic system is unanimously regarded as a fully consolidated regime, widely recognized and legitimized by all political, economic and social forces. I have strived, throughout my fieldwork and research for this study, to be cautious against establishing an absolute correspondence between that history of dictatorship and present reactions to current legislative measures regarding the implementation of video surveillance in the public space, finding for example, that an experience of past repression can as easily reveal an instinctive resistance to such measures, as a readiness to accept them. Nevertheless I believe, based on my interviews and the study of specialized literature on the Portuguese dictatorship and its repercussions, that there is at least one point on which there are few doubts: the democratic system in Portugal is generally perceived as irreversible by all sectors of Portuguese society, including all relevant political forces, the military and justice institutions, police forces and public organisms such as the Data Protection Authority, as well as by the vast majority of citizens.

There is therefore absolutely no credible threat to its existence as granted by the Portuguese Constitution. But of course, this does not prevent voices immediately being raised at even the slightest hint of securitarianism in political discourse, connoting it with an authoritarian past.

Backwardness as a Syndrome

Before starting a political characterization of the past century, it is important to stress once again that although the dictatorial regime in Portugal corresponds to a very specific time period contained within the last century, the fact is that throughout its eight hundred years of national history, Portugal has been dominated for the most part by a monarchic regime. This was only interrupted for a brief interlude during the liberal revolutions of the nineteenth century when it was substituted by the short-lived Republic of 1910, which ended in 1928 with the instauration of a military dictatorship. It should therefore always be kept in mind, especially when considering what I call the successive waves of modernization that occurred over the last century, that the fact that Portugal has mostly alternated between absolutist regimes and cults of personality – that is, regimes that exalt the individual qualities of its leader – unquestionably marks Portuguese mentality and its social life.

As mentioned above, the feelings of national pessimism and backwardness that find frequent expression in political and common discourses have a long history in this country. Its symptoms have permeated the philosophical, artistic and academic spheres since the nineteenth century, corresponding to a period of severe economic decline. Running parallel to this fatalism in Portugal there can also be found a strong belief in particular traits – such as courage and resilience – that are given a national dimension akin to the notion of a 'Portuguese spirit', which usually takes some of the country's remarkable achievements – most notably those connected to the maritime conquests of the fifteenth and sixteenth centuries – to be its direct embodiment. This contradiction is often resolved by attributing to some unknown or mysterious forces the responsibility for both the setbacks that the people are continuously subjected to as well as their determination to continue overcoming them.

Among the works that most expressively portray this spirit of self-criticism and resignation are those of Eça de Queiroz, whose relevance as a social and political satirist is attested by the fact that his literary novels and essays provide a vivid depiction of this historical period while maintaining all their actuality even today. The works of philosopher and essayist Manuel de Laranjeira, who ended his own life at the beginning of the twentieth century, must also be noted for their direct references to the so-called 'national pessimism' and the disenchanted reflections on the failure of the republican regime.[1] More recently, the philosopher Eduardo Lourenço has also dedicated particular attention to the question of the Portuguese character, most notably in his work of 1978 *The Labyrinth of Saudade – Mythical Psychoanalysis of Portuguese Destiny*,[2] where he discusses precisely the feeling of fatalism and anxiety over the return of past grandeur. Among contemporary authors there is also the philosopher José Gil who identifies persisting traces of dictatorship today – for instance in the general passiveness and weak civic awareness that is reflected in the absence of an active participation of the common citizen in social affairs.

The sociologist Boaventura Sousa Santos, reflecting on the role played by Portuguese historical legacy on the economic, political and social spheres, ventures to ask, 'How much does the past weigh in on the present and future of Portugal?' (2009: 1). In a few words, the author claims that in this case the diagnosis itself has become part of the problem instead of its solution, helping to perpetuate both the idea of the country's backwardness and the very arguments which supposedly justify that condition. What this ultimately implies is that modernization ends up being turned into a myth that reflects a blind faith rather than an attainable ideal, much like the following remark on Africa by Ferguson: 'the myth of modernization (no less than any other myth) gives form to an understanding of the world, providing a set of categories and premises that continue to shape people's experiences and interpretations of their lives' (Ferguson 2006: 14).

An answer to the question posed by Boaventura Sousa Santos could be that as far as video surveillance is concerned, the past still holds a heavy, though subliminal, weight on present common discourse, and therefore, despite not threatening the existence of democratic rule in Portugal, marks the balance of power involved in institutional relations and thus conditions the modus vivendi of the Portuguese democratic system. As will be seen, the establishment of a system based on state scholar corporatism, the creation of the political police, the outbreak of a colonial war on multiple fronts (during the decade of 1960) and the vast fluxes of emigration are among the factors that contributed to render Portugal increasingly obsolete, doing nothing to bring change to its mostly rural and uneducated population during most of the twentieth century. The end of the dictatorship in 1974 brought with it a newfound impetus of 'political modernization' that manages to contain extremist reactions on either side of the political spectrum and establish a multi-party democracy, while bringing Portuguese colonial aspirations to a definite end. Above all, the end of the Estado Novo was seen as an opportunity to carry out a whole set of social reforms that would improve standards of living and the instauration of democratic rights and freedoms.

In the second half of the 1980s the structural changes effected by Portugal's entry, in 1986, into the European Economic Community (today's European Union) led the way for an internal project of what I call a period of structural modernization. The development of the democratic transition that started in the mid-1970s reached its apex or point of definite consolidation during the following decade and into the 1990s. At this point Portugal had shortened its distance from European countries considered more advanced, finally managing to attenuate marked divergences in terms of health, education, housing and employment policies, amongst others. Moreover, this new stage paved the way for a new competitive economy within the broader project of the European Union, and with it, the chance for a much hoped-for place in Europe.

Between the end of the 1990s and the beginning of the twenty-first century, the country went through a cycle of economic recession (the first serious test of its recent European journey) with the effect of exposing structural fragilities that still existed. This fragility was reflected on the political stage by the succession of governments from both centre-block parties that did not reach the end of their term: the socialist Prime Minister António Guterres renounced his office halfway through his second term after the party was heavily defeated in the mid-term local elections (his terms of office were 1995–1999 and 2000–2002); and the social democrat Prime Minister Durão Barroso was elected in 2002 but abandoned power in 2004 to accept an invitation to preside over the European Commission.

Just as the use of the term 'modernization' in this book is the appropriation of a category applied by my interlocutors, belonging to the domain of collective representations and conveyed equally in political, media and common discourses, so too the following identification of its different stages is based on a purely personal interpretation. It is therefore not supported by any of the common concepts used by other social scientists, who may rightly find objections to this terminology, namely those who may consider that none of these stages was effectively accomplished. For authors such as António Barreto or Boaventura Sousa Santos, Portuguese modernization has never actually come into full existence, inasmuch as it has always been marked by the chronic backwardness inherited from a mindset that has ruled the country for decades, if not centuries. For them, modernization in Portugal is forever bound by atavism. Nevertheless, it is precisely this duality that this study intends to address, in the sense that by considering the changes which have actually taken place, and unquestionably so, it is situating it within a historical record that is not erased, but rather incorporating their underlying motives.

This phenomenon applies equally to other southern European states such as Greece and Spain, whose political histories closely resemble the Portuguese case (Bermeo 1997; Morlino 2010; Pinto 2010; Palacios Cerezales 2010). In both these countries, dictatorships based on repressive policies that used methods of control and identification of citizens as a means to impose political and moral values, explicitly conservative and religious ones, continues to leave a mark on social life regardless of their current stage of development. Just as in Portugal, these countries have undergone successive reforms with a view to gaining their place in Europe and thus becoming economically competitive powers in their own right. Nevertheless, these efforts have not wholly eradicated the old mentality and modus operandi of their historical legacies, which continue, as will be argued in the Portuguese context, to produce effects on present cultural and popular imaginations.

The next section will consider some of the significant moments in the Portuguese history of the twentieth century, beginning with a description of how the policies of authoritarian repression begun in 1933 shaped a pathway to progressive isolation and led to the decay of socio-economic conditions in Portugal in comparison to other western European states.

Political Modernization: Salazar's Estado Novo and the Carnation Revolution

António Oliveira Salazar's gradual rise to power, beginning in 1928 – first as Minister of Finance in Marshall Carmona's government, and later as head of government, with the popular approval in 1933 of the constitution that laid the base for his so-called Estado Novo – would only end in 1974, lasting roughly four decades. The regime, basically an embodiment of the values contained in the triangle God–Nation–Family, translated into a strategy that progressively led to the isolation of Portugal from the rest of Europe, in favour of a political and economic project that insisted in maintaining a colonial empire in countries such as Guinea-Bissau, Angola, Mozambique, East Timor and São Tomé and Príncipe.

The essentially conservative and Catholic direction of Estado Novo's political and economic project was reflected internally in an ideology of power that extended to all social spheres, enforcing its overwhelming authority through such means as political and cultural censorship (most notoriously control over all media and communication channels); the increasingly pervasive interference of a political police (the PIDE) and the promotion of rural subsistence and state corporatism. As Costa Pinto (2003) describes it, the Estado Novo's political regime inherited the previous military dictatorship's repressive machinery, adding to it the mechanisms of propaganda and intelligence and making its organs increasingly more dependent on the direct control of the single authority figure of Salazar, whose role as leader of the nation gradually assumed an almost messianic character, as the country's 'saviour'. The whole public administrative apparatus was centralized in the hands of the regime, which notwithstanding some essential differences, to a certain degree followed the contemporary political trends of other systems, like the ones instituted by other European leaders such as Franco's military dictatorship in Spain, or even the fascist states of Mussolini and Hitler.

Salazar, basically a traditionalist both ideologically and culturally, had an absolutely anti-liberal view of state affairs, which he considered should be put at the service of a political project based on the philosophical identification of religious and nationalistic values. His total rejection of democratic society was supported by his organic conception of the state, which implied resistance to all the advances of modernization, considered more as threats than as models. His idea of popular education was closely linked to a system of indoctrination, breeding the exaltation and revival of the nation's past grandeur, its conquests and colonial riches, thus casting out all values that were contrary to the regime, and otherwise smothering ideological open-mindedness (Pinto 2003).

The Salazarist regime, despite its authoritarian quality, cannot be completely identified with the fascist model imposed in countries like Germany and Italy. It is true that the Estado Novo rejected the democratic model, and even though theoretically other political parties were never legally prohibited, in practice it established a single-party system in many respects similar to other European dictatorships. Nevertheless, the National Union, conceived as a kind of 'national front', never fully adopted the totalitarian philosophy of unrestrained power above everything and everyone (Lucena 1994). Therefore, despite invariably constituting an autocracy based on a political rhetoric that defended the supremacy of the state, considered the touchstone of political order, this type of authoritarianism 'did in fact admit, to a greater or lesser degree, limitations on the powers of the state, conceding some measure of autonomy to other social bodies' (Morgado 2010: 12).

In any case, obedience and submission to the regime would in time imply an ever more extensive network of its intelligence organs, through the dissemination of the PIDE throughout the whole Portuguese territory – in which the so-called ultra-maritime regions were included. This involved the creation of prison and deportation camps specifically intended for political opponents, such as the Tarrafal Fortress in Cape Verde (Mateus 2004). The primary mission of the political police was to detect and neutralize political opposition, mainly communist and anarchist activists, and according to Fernando Rosas, can be said to constitute the backbone of the regime. This organ's action relied on a powerful apparatus of surveillance, through a system of information gathering and investigation that allowed it to carry out persecutions and political cleansing. This activity served the purpose of perpetuating the regime's longevity through 'civil demobilization, instilling fear, subservience and generalized intimidation' (Rosas 1989).

The significance of the PIDE in Portugal, in its double role of indoctrinating and controlling information on citizens and their movements – making it one of the regime's most important organs – has been thoroughly studied by Irene Flunser Pimentel, who stresses the fact that the theme still holds a 'traumatic dimension for many of its survivors' (Pimentel 2007: 13). This traumatic memory, mostly felt among older generations, closely resembles the vivid description given by Samatas of the Greek case:

> Greece's authoritarian surveillance culture casts a long shadow. On the one hand, the lengthy tradition of surveillance continues to allow state and private organizations to commit extensive privacy violations and abuse of minority rights. This contributes to lingering sensitivities amongst Greeks – especially the

> older generations – about forms of surveillance in the aid of political control. The painful memories retained from the decimations suffered from official discriminative 'filing' reinforce a mutual mistrust in state–citizen relations. (Samatas 2005: 188)

Besides the instruments of censorship and police control, many other organisms were created in Portugal to ensure the upkeep of public order and security, whose structure was based on the (in some cases compulsory) recruitment of citizens and whose activity relied on the active participation of its members. The FNAT (Fundação Nacional para Alegria no Trabalho, or National Foundation for Joy through Work) was a recreational institution similar to its Italian and German counterparts the OND (Opera Nazionale Dopolavoro) and KdF (Kraft durch Freude), whose main goal was to occupy workers' free time while instilling in them the regime's doctrine and practice. The Mocidade Portuguesa (Portuguese Youth Movement) was an organization for young men and women, enrolment in which was compulsory for all children between the ages of seven and fourteen years (Arriaga 1976; Pimentel 2000); and the Legião Portuguesa (Portuguese Legion) was a paramilitary organization whose mission was to 'defend the Nation's spiritual patrimony and fight the communist and anarchist threats' (Madeira, Farinha and Pimentel 2007). Analogous Italian and German counterparts are found for any of these institutions, in all cases intended to promote submission and obedience to the respective regimes in power (Khazanov and Payne 2009).

The policy of state corporatism on which the regime was rooted further enhanced its authoritarian character, enabling it to control the country's economic activities and determine working conditions. The Estado Novo's value system thus created 'an apparatus of authoritarian and statist ideological inculcation embedded in people's everyday lives (within the family, school and recreational activities), which crossed all spheres of social life, seeking to mold society after the image of its head of state' (Rosas 2001: 1031).

As late as the 1960s, Portugal was still a country whose population could be described as mostly illiterate, and it faced a colonial war – the so-called Guerra do Ultramar – that would add to the already difficult social and economic situation. This was reflected in the vast waves of precarious (often clandestine) emigration to countries such as France, the United States of America, Brazil, Canada, Luxemburg or Venezuela. During this decade, almost 12 per cent of the population was in one way or another forced to leave the country, which meant that 170 000 people had to look abroad for better conditions or merely to escape the colonial war or political persecution (Barreto 1994). With its back increasingly turned to Europe and the world, this regime's isolationism – actually assumed in its dictum 'Proudly alone' – only increased as other major western economic powers in Europe gradually abandoned colonial interests and turned away from models of political authoritarianism, leaving southern countries such as Spain and Portugal in an ever more fragile position.

The Estado Novo regime suffered yet another important blow when in 1970 it lost its figurehead, as Salazar succumbed to medical problems which had debilitated his health in the previous couple of years, more precisely since he suffered an accidental fall at his summer house in August of 1968. The strong personality cult around which the regime had been erected was to prove decisive in hampering the chances for successful transition to his successor, Marcelo Caetano, resulting in the system's relatively swift breakdown once that central element had been removed.

It was within the context of this generally impoverished country of rural and patriarchal customs that the military coup led by the Armed Forces Movement on 25 April 1974 finally put an end to the dictatorship. The process of democratic transition, which became known as PREC (Processo Revolucionário em Curso, or Ongoing Revolutionary Process), was initially marked by some instability, giving rise to strong popular and political demonstrations, and instances of extreme behaviour on both sides of the political spectrum that included political cleansing, or the attempted coups of left- and right-wing extremists (in March and November of 1975). The decolonization process was also initiated in this period, forcing whole generations of Portuguese citizens to return from the colonies. The Preamble[3] to the Constitution of 1976, which has remained unchanged until today despite the changes that have been made to the rest of the text since then, is clearly demonstrative of the urgency and motivation that underlie this democratic process.

> On the 25th of April 1974 the Armed Forces Movement crowned the long resistance and reflected the deepest feelings of the Portuguese people by overthrowing the fascist regime.
>
> Freeing Portugal from dictatorship, oppression and colonialism represented a revolutionary change and the beginning of an historic turning point for Portuguese society.
>
> The Revolution restored their fundamental rights and freedoms to the people of Portugal. In the exercise of those rights and freedoms, the people's legitimate representatives are gathered to draw up a Constitution that matches the country's aspirations.
>
> The Constituent Assembly affirms the Portuguese people's decision to defend national independence, guarantee citizens' fundamental rights, establish the basic principles of democracy, ensure the primacy of a democratic state based on the rule of law and open up a path towards a socialist society,[4] with respect for the will of the Portuguese people and with a view to the construction of a country that is freer, more just and more fraternal.

The transitional stage to the consolidation of political democracy lasted from 1974 to 1976, the year in which the new Constitution was ratified and the first presidential, legislative and local democratic elections were carried out. General Ramalho Eanes was elected president and Mário Soares, recently returned from exile, headed the government won by the socialist party's first electoral victory. He would be succeeded in that post by

Jorge Sampaio who, like Soares, had also been actively involved in the anti-Salazarist struggle. In the words of J.M. Ferreira:

> the electoral preponderance of the socialist party since the first elections for the Constituent Assembly would determine the establishment of a pluralist regime of political democracy. . . . From that date on, up until 1982, right-wing authoritarian tendencies have been subdued and marginalized by the successive leaders responsible for the armed forces' political strategy. (Ferreira 2001: 16)

Even the presence of right-wing parties was closely controlled until the 1990s in Portugal's young multi-party democracy, a clear sign of the country's political experience – also a policy of memory – which was more accentuated in the decade immediately following the events of 25 April 1974 up until the 1980s when Portugal was admitted into the EEC.

It was a time of great political turbulence and above all of popular liberation, in which people purged themselves of the Salazarist ideals – namely religious and cultural conservatism – but also worked to effect the deep transformations that Portugal needed in terms of establishing a social state based on more equitable health and education systems and improving working conditions through the creation of syndicates. The lyrics to the song 'Freedom' by Sérgio Godinho, one of the most popular singer-songwriters to come out of the democratic revolution, expressively convey this double mark of past experience and future expectations in the collective imagination of Portuguese representations.

> We have carried the weight of the past and the seed
> Waiting for so long has only increased the need
> Only a torrent will quench the thirst of delay
> We have lived so many years in a whispering hall
> You can only want everything when you've had nothing at all
> You wish for a full life when it has been just one big hole
> Freedom will only come when we find
> Peace, bread
> A roof over our head
> Health care
> Education
> Freedom will only come when we find
> Freedom to change and decide
> And the people own what the people produce
> – Sérgio Godinho, 'Freedom' (1974)

Barreto accounts for this process by explaining that in the months immediately after the military coup,

> with the foundation of the democratic state, there were pressures from the whole society to democratize all community/collective activities, not just political life. With relative swiftness, the old ways of paternalism, bureaucratic despotism,

> secrecy in public administration, social segregation and favoritism all declined. Everyone became acquainted with their rights and equality, even before the law had been guaranteed. (Barreto 2003: 179)

In terms of social equality, some of the most significant transformations were connected with the rights of women, who were no longer expected to live subjugated by the authority figures of their fathers or husbands. The rights to divorce and civil marriage also began the process of rupture with the past (Mateus 2001; Salgado Matos 2001).

Within the historical frame of the past century, this particular period corresponds to what could be called a project of political modernization. The transformations that were effected in the political and social spheres during this time have left an especially deep and enduring mark on the country's development up to the present date. This is true for all spheres of Portuguese life, but even more so in terms of what may be referred to as the 'weight of the past' on the collective imagination. This is because its common evocation and invocation in today's general discourse occur independently of citizens' individual and direct experiences of the repressive and authoritarian system, which in fact are limited to the older generations. This period symbolizes a form of existence which no one wishes to go back to, and nevertheless is still occasionally revived, as was seen in the reactions caused by the political campaign that proposed the implementation of public video surveillance for Lisbon. But more than a mere memory, the truth is that a great number of institutional agents, and particularly police forces, claim to feel that their actions are still conditioned precisely by the negative charge left by the period of dictatorship.

The Country's Modernization: Entry into the EEC and Structural Reforms

The period between 1976 and 1986 was characterized by a political stability that was marked by the negotiations leading up to Portugal's admittance into the EEC, followed by the process of 'Europeanization' of Portuguese democracy. As Sousa Santos and Nunes observe, 'the "normalization" that followed the revolutionary period led to the establishment of a parliamentary democratic regime and to the drafting of a Constitution which tried to inscribe both the institutional framework of parliamentary and representative democracy and the innovative forms of participatory democracy' (Sousa Santos and Nunes 2004: 11). During this decade, the EEC was considered a symbol of democracy and a way out of Portugal's past isolation. It represented the entry gate into a new world economy that would bring commercial investment and social progress and generally develop a country which had stagnated in almost all areas (Pinto and Teixeira 2002; Viegas and Faria 2009).

When Portugal finally became a full member of the EEC in 1986, it was tantamount to conquering a whole new level of economic, political and social development for its citizens. It meant not just the international confirmation of this democratic system's consolidation, but additionally a boost for foreign relations and investment, as well as commercial exchange and corporate partnerships. Overall, it was the fulfilment of long held expectations.[5]

As in other countries undergoing similar political transitions during this decade, the changes in Portugal were both swift and extensive (even though, as will be shown, their results often fell short of European standards):[6] in the spheres of industry, education, health, justice and social security – which relied on the significant role played by a welfare state system – the country's outlook, radically changed in the direction of progress and improved living conditions, was transformed at a rapid pace. Pinto and Teixeira go so far as to claim that:

> Democratic consolidation, EEC accession and economic development all coincided to create a virtuous circle that could not have been foreseen at the moment of application. To the surprise of many sectors of public opinion, in 1990 Portugal lost its status as an 'under-developed country', a label that had been used to characterize the country ever since the concept had been devised. . . . Rather than the catastrophic prospect that seemed to loom large for Portugal during the 1970s, the country managed to consolidate its democracy and to take important strides forward in its social and economic modernization as a member of the European Union. (Pinto and Teixeira 2002: 36)

Ironically, the initial period of enthusiastic civic and popular participation witnessed during the previous decade gradually waned as the decade of Europeanization progressed, and was substituted by a decreasing interest on the part of the population at all levels of public activism. A general numbness to the country's political direction seemed to invade all spheres of civilian life, regarding both internal policies and foreign affairs, particularly those having to do with Portugal's position within the EEC/EU. This trend assumed numberless forms and is revealed, for instance, in general state policy, which was pushed towards an indistinct centrism, in which the alternation between right- and left-wing governments seems to have become a purely perfunctory chair-swap.

The memory of the dictatorships' legacy receded from people's daily experience, as did revolutionary ideals. Several historians and political scientists observed that civic mobilization was registered more intensely during the first ten years after the fall of the old regime, to be followed by a growing indifference for collective political life. One of the most telling signs of this apathy can be found in the gradual rise of abstention rates in the elections carried out between 1980 and 1990, reaching between 40 per cent and 50 per cent (Barreto 2002; Magalhães 2001, 2003, 2008).[7]

António Barreto demonstrates that in the years between 1985 and 1995, the decade of full European integration, certain economic and social problems were intensified, and even though Portugal did not revert to its previous condition of precariousness and underdevelopment, they reveal the structural and economical fragilities of this peripheral country. The visible structural modernization that has occurred, and that seems in all respects to place Portugal within the standards of other European states – those that represent precisely this model of economical, technological and political progress – is, however, apparently unable to generate those basic resources that are most essential for the system's sustainable upkeep. I am referring of course to the acquisition of the skills that are essential to make the country competitive. These changes can be observed in Table 1.1, which represents the changes that occurred in the most significant indicators of sustainability between 1960 and 2011.

This period can be split into two different stages: an initial moment of European integration, which lasted for the second half of the 1980s, in which high growth levels were registered in all areas of Portuguese economic, political and social life; and a second stage that occupied the first half of the 1990s, in which this growth s quickly stalled. Once again, examining the developments that took place in Portugal between the 1960s and 1990s, Barreto explains:

> In the Portuguese society, which opened up to the world through emigration, through the channels of mass media, through European integration and market globalization, all expectations are permitted, both in the domain of consumption and economical promotion, as in those of social mobility and protection or security. . . . It so happens that material resources, economic and entrepreneurial means, productivity, technological and scientific capabilities and competitive experience fall below the overwhelming majority of other western countries with which Portugal has the closest bonds. (Barreto 1995: 851)

In short, after a period of euphoria and enchantment, the frailties of an economy that had been expanding at an unbridled pace, fuelled in great part by the inflow of structural and cohesion funds, determine the gradual slowdown of the country's actual growth rates by the end of the 1990s. This phenomenon exposed the reality of a divergence between real economic growth and the increase of consumption and public investment, between spending money and the strengthening of production and productivity. A growth based on consumer and public debt that was not matched by real productive capacity and national income ended up jeopardizing (once again) the very process of modernization and development.

Added to the internal economic crisis, a new cycle of political crisis began to install itself in the first years of the twenty-first century, further compromising the whole structural consolidation project. As mentioned at the beginning of this chapter, the repeated failure of governments to reach the end of their legislature, whether due to prime ministers abandoning their posts – as in the cases of Guterres in 2001 and Barroso in 2004 – or more revealingly, to the frustrated government of Prime Minister Santana Lopes'

Table 1.1: *Portuguese Society in Key Figures* (Summary Table)

	1960	1970	1981	1991	2001	2011
Total resident population	8 889 392	8 611 125	9 833 014	9 867 147	10 356 117	10 555 853
Crude death rate (‰)	10.7	10.7	9.7	10.4	10.2	
Infant mortality rate (‰)	77.5	55.5	21.8	10.8	5	
Crude birth rate (‰)	24.1	20.8	15.4	11.7	11	
% of resident population aged 15 and over that cannot read or write	65.6	35.2	21.9	13	9.2	
% of resident population aged 20 and over with higher education qualifications	0.9	0.9	2.4	4	7.9	
Graduates	x	x	x	18 671	61 140	
Ph.D.s concluded	x	61	112	319	908	
Doctors per 100 000 inhabitants	79.8	94	197.9	284.2	322.9	
Births in health establishments as a % of total	18.4	37.5	73.8	99.4	99.7	
Judges per 100 000 inhabitants	2.8	4.9	7	10.3	13.9	
Lawyers per 100 000 inhabitants	22.1	29.7	46	95.6	183.5	
Actual prison occupancy (prisoners/prison capacity x100)	77.4	47.2	75.7	108.4	114.5	
Activity rate (% active population/population aged 15 and over)	x	x	64.4	63.7	61.5	
Number of unemployed persons (thousands)	x	x	165	208	214	
Number of museums, zoos, botanical gardens and aquariums	99	122	121	320	234	
Number of museums', zoos', botanical gardens' and aquariums' visitors (thousands)	1313	2750	2980	6578	8556	

Number of contributor employees – Social Security – general system	78 004	118 530	420 756	501 925	628 889
Pensioners – Social Security	56 296	187 297	1 719 685	2 230 326	2 528 926
Average annual pension (euros – constant prices 2006) (expenditure on pensions/total number of pensioners) – Social Security	890.80	1410.10	1715.60	2452.80	3515.90
Annual social security expenditure (thousand euros – constant prices 2006)	345 478	1 431 316	5 339 409	9 424 802	15 905 927
Government expenditure (final budget, million euros – constant prices 2006)	3189.1	6575.4	20 760.9	41 789.3	55 586,1
Government revenue (budget application, million euros – constant prices 2006)	3218.5	6793.4	20 762.6	41 789.7	52 339.8
GDP (thousand euros – constant prices 2006)	27 765 135.2	48 694 930.8	80 324 252.9	116 082 480.4	154 758 286.1
GDP per capita	55.4	134.9	992.8	6389.0	13 031.9

x – there are no data for these years.
Source: Pordata (http://www.pordata.pt)

short-lived career, brings an atmosphere of instability to political life. The latter case is especially notorious because it was the first time since the Estado Novo that a prime minister assumed office without previous popular election, and the president was called upon to intervene by dissolving it and calling new elections, something which was also unprecedented in Portugal's recent democratic history. Once again, the country seemed to desperately await the arrival of a strong leader, someone to single-handedly assume the country's destiny and rekindle the promise of modernity.

Fighting Backwardness through Technology: The Sócrates Era

After having been Minister for Environmental Affairs in the second socialist party government of António Guterres, José Sócrates emerged as prime minister in this context of political and economic turmoil, initiating Portugal's seventeenth constitutional government in February of 2005. Sócrates, an engineer by profession, managed to secure this victory in a political campaign that set him against a renowned economist and long-time member of the social democratic party, Manuela Ferreira Leite. While the latter's campaign set out to warn the electorate of the need for a change in direction towards more severe austerity measures and budget control, José Sócrates on the other hand presented an altogether different approach, which exuded an optimism bent on emphasizing national virtues and the country's promising potential. He proposed to put the country back on the track of international competitiveness by means of what he advertised in his campaign as a 'technological revolution', bringing the idea of modernization once again to the forefront of the political agenda.

Sócrates can be said to have initiated a project of technological modernization, one which is here interpreted as the continuation, or renewal, of a modernization design that has already been identified as a recurring theme in this Portugal's recent political history. The last decades of Portuguese democracy can in fact be divided into corresponding stages of modernization, each with its own characteristic approach. While the end of the 1970s and beginning of the 1980s witnessed a phase of political modernization, and the years between 1985 and 1995 an era of structural reform and economic modernization, the best characterization for the period between 2005 and 2011 is to be found in Sócrates' campaign slogan, 'modernization through technology'.

This positive approach proved to be an excellent strategy, at least in electoral terms, and may have been decisive in giving Sócrates an overwhelming victory, with 45 per cent of the votes, compared to 28 per cent conquered by his direct opponent. Although he would be reelected in 2009, the second time around results would be far less expressive, making Sócrates lose the majority in Parliament that had been so important during his first mandate, and eventually costing him his place in 2011.

The Technological Plan presented below provides a clear instance of how the incentives of competitiveness and modernization can be used as an argument – as it has been at a global scale – to implement political measures and condition strategic decision-making. In fact, the elements of competitiveness, innovation and technological development have always been hailed as the beacons of progress and advancement, namely in the EU's strategic plan for an expanding 'Europe'. As Gledhill explains in his description of Foucault's analysis of strategies, technologies and programmes of power, these 'define a domain of social reality to be turned into an object of rational knowledge, intervened in and made functional'. According to this author:

> Technologies of power are techniques and practices for the disciplining, surveillance, administration and shaping of human individuals. Programmes define forms of knowledge and discourses about objects of knowledge. Technologies are apparatuses of power designed to implement that knowledge. Strategies of power are what agencies do in practice in exercising power and in operationalizing programmes and technologies. (Gledhill 2000: 150)

This kind of rationale is evident in the following transcript of the introductory text to Sócrates' Technological Plan:

> The Technological Plan is an action agenda for all the Portuguese society, which aims at mobilizing enterprises, families and institutions for surpassing the modernization challenges the country has been facing during the last years. Within this context, the Portuguese Government has assumed the Technological Plan as a priority in the implementation of its public policies.
>
> Besides, the measures aggregated in the Technological Plan constitute the pillar for Growth and Competitiveness of the Portuguese National Reform Plan, designated National Action Programme for Growth and Jobs 2005–2008.
>
> As a part of the governing programme, already approved of by parliamentary vote, the starting point of the execution of the Technological Plan has coincided with the beginning the XVII Constitutional Government in Portugal. . . .
>
> As a strategy meant to promote the development and reinforcing of growth and competitiveness in Portugal, the Technological Plan is based on three axes:
>
> Knowledge – To qualify the Portuguese for the knowledge society, fostering structural measures which aim at enhancing the average qualification level of the population, implementing a broad and diversified lifelong learning system and mobilizing the Portuguese for the Information Society.
>
> Technology – To overcome the scientific and technological gap, reinforcing public and private scientific and technological competences and recognizing the role played by enterprises in the process of creation of qualified jobs and Research & Development (R&D) related activities.
>
> Innovation – To boost Innovation, helping the productive chain to get adapted to the challenges of Globalization by means of the diffusion and development of new procedures, organizational systems, services and goods.[8]

It is obvious that the previously mentioned operative concepts of advanced societies are made exceedingly clear in the Technological Plan, which, as one can read, was 'part of the governing programme' conceived as an 'action agenda for all the Portuguese society'. This agenda would include a multitude of other initiatives, of which Simplex (a catchy name that was chosen for the administrative and legislative simplification programme), as the paragon of a policy aimed at increasing the transparency, agility and expediency of bureaucratic procedures, is a clear example. But more significantly, all these programmes are linked by the common denominator of technological innovation; that is, the introduction of informatics systems and digital technology in the most diverse spheres, intended not only to modernize procedures and modus operandi working methods, but ultimately also to transform mentalities.

On this point the Portuguese case presents a significant feature: modernization here is not restricted to technological innovation, but also implies popular conceptions of what it means to be 'modern', almost in the sense of fashionable or sophisticated. Sócrates' political success was probably due as much to this appealing feature as to any of its other more substantial merits. Once more, the need for comparison with other countries, of 'borrowing' foreign models as the only possible criteria, seemed to haunt the Portuguese imagination.

Only a macro-historical, macro-political and macro-economic analysis can actually provide a full understanding of the meaning and scope of the technological measures that have been emphasized throughout this chapter. The fact is that at the turn of the twenty-first century, the whole governmental strategy for development has, in one way or another, relied on the idea of using technology as its primary vehicle. Within this strategy, governmental measures have taken technology's instrumental innovativeness and raised it to a symbol of modernity.

While in other countries the effects of 9/11 have greatly influenced the debate surrounding issues of surveillance, personal identification and privacy, incontrovertibly linking it to the global policies of a war on terror, such considerations cannot be said to have the same bearing on the context of the technological-minded Portuguese 'war on backwardness', beyond the prominent role played by technology in either case. The war, in this case, was of a completely different kind; it was a war against underdevelopment, bureaucracy and ineffectiveness, to be fought with the ultimate modernizing weapon – technology.

One proof of this is that the implementation of public video surveillance in Portugal did not emerge within a broader strategy for the strengthening of national security or defence. It arose within the strict logic of urban rehabilitation and crime fighting. Nevertheless, the warfare analogy can be extended to one other aspect, connected with the shock value attached to this particular form of intervention. Just as the U.S. military doctrine of 'shock and awe' is based on the idea of rapid dominance, so too the Portuguese technological plan was intended to have a rapid and overwhelming effect, albeit in this case not as a military action pursuing geopolitical goals, but as an administrative reform of national mentalities and procedures. The Technological Plan announced by the Sócrates administration was indeed referred to as a strategy of 'Technological Shock'.[9] The metaphorical appropriation of military vocabulary is apparently a feature of other developmental government programmes, as observed by Ferguson: 'Military metaphors – successful "operations", well-executed "attacks" and "assaults" on "targets", etc. – abound, and the image is of a single, unitary, righteous army doing battle against a universal enemy' (Ferguson 1994: 65).

In any case, it can be said that within this model of technological modernization, the element of shock – ultimately intended to spur on Portugal's competitiveness and raise the country to neighbouring countries' standards – was unquestionably one of the essential elements in the policies adopted from 2005 onwards, during the socialist party's governance. In another sense, it might be suggested that the very idea of competitiveness was infused with a comparative logic, in that it was constructed on the basis of comparison to models that are always ultimately considered external.

The downside of a strategy that relies on looking up to the 'other' (countries, citizens, lifestyles, living conditions etc.) is that its results also become inextricably bound up with those same external references. European integration brought with it the promise of innovation and mentality change, but the hard fact is that the structural reforms needed to increase economic resilience, bring dynamism to civil society and raise levels of training and expertise did not keep up with the rhythm of those intentions. It thus had the inverse effect of further accentuating Portugal's peripheral condition within the EU, as it failed to respond to the standards set by this community of countries in most spheres of action. As the philosopher José Gil explains:

> In these first years of the twenty-first century we witness the desperate effort for Portugal's self-assertion; that is, to ensure Portugal's survival to its inclusion in Europe . . . [. W]e seem less interested, for instance, in promoting exchange, interconnection, blending, osmosis, permanent channels of communication with other European cultures, than in being assured that our products are properly validated abroad. This 'abroad' remains a remote place to us. (Gil 2004: 63)

In a certain sense, being peripheral is not just an actual condition, but persists as a meta-discourse that cyclically reifies the national syndrome of 'being' Portuguese; that is, of being underdeveloped and backward. Of course, the greatest deception is to think that this Europe that Portugal sees itself as trailing behind is immune to those same problems. Moreover, it prevents internal debate from establishing points of contact with other countries in the region, namely other southern European countries with similar political histories, and realizing that Portugal shares many common features with them.

Returning to the question put by Boaventura Sousa Santos regarding the weight of the past on the present and future of Portugal, it is clear that, similarly to other peripheral or semi-peripheral countries, Portuguese history over the past century shows that

> the problem of the past tends to be more serious given: a) the more widespread the social perception of the distance between the country in question and the more developed countries that serve as models; b) the more hegemonic the representation that such distance could and should be smaller; c) the more credible the idea that this has not been possible for internal causes or conditions. (Sousa Santos 2009 399)

The fact that a peripheral country such as Portugal is just taking its first steps in the use of video surveillance in public areas constitutes a great opportunity to understand how such a process is put into practice and how the surveillance is subsequently sustained. The chain of events that forms Portugal's recent history – Salazar's dictatorship, the 1974 revolution, the democratic transition and later inclusion in the EEC in the 1980s – might reflect an evolution that seems to point in a very clear direction. This appearance of linearity makes it very tempting to jump to conclusions when interpreting the country's present framework, in terms of issues such as invasion of privacy, public monitoring or state control.

However, once again, deciphering what the recent policies and public opinion actually reveal about the country's present context is, as will be shown, considerably more problematic. The main difficulty resides in the fact that the effects of past experiences tend to manifest more deeply within the sphere of mentalities than in the projects themselves, even when the projects that seek to implement public video surveillance can be connoted with authoritarian regimes.

Therefore, claiming that in Portugal the quest for development and modernization has been a constant in successive historical periods – and it should once again be emphasized that this does not only hold true for the past century – is tantamount to reifying its persistent backwardness. And it could be argued that this dichotomy is revealed in the forms of interaction assumed by the different spheres of power involved in a process that evidently does not reflect a common goal. António Barreto referred to the modernization process as an 'avalanche'[10] taking everything in its way, destroying much of what is built but leaving a few remnants behind it. This metaphor alludes to the fact despite the many transformations undergone by Portuguese society over the last thirty years and eloquently represented in Table 1.1, systems and outlooks which for lack of a better word can be called archaic continue to live side by side with more modern methods and procedures. This double condition carries with it yet other intrinsic phenomena: the resistance to change, the suspicion of declared intentions, and paradoxically at the same time the fascination and obsession with the promise of things to come.

Notes

A previous version of parts of this chapter is published in Frois (2012).

1 Identical reflections and analyses have been provided by Castañares and Juste for the Spanish case. 'Terms such as failure, backwardness or decadence define

the evolution of the Spanish people . . . [and] throughout Spain's history, there appears to have existed three major paradigms: that of decadence, with its roots in the 17th century; the romanticism of the nineteenth century, which sees Spain as the country of drama and violence; finally the paradigm of exceptionality that gathered force during the twentieth century' (Castañares and Juste 2002: 42). The authors also refer to several intellectuals and scholars who have reflected on the topic, namely José Ortega y Gasset, José Antonio Maraval, Pedro Laín, Julían Marias, Salvador de Madariaga and Francisco Avala.

2 In Portuguese, *O Labirinto da Saudade* (Lourenço 2001). The works of Eduardo Lourenço are translated into several languages. The cited essay is included in English in *Chaos and Splendor & Other Essays* (Lourenço 2002).

3 This transcription was taken from the seventh version of Portugal's Constitution, published in 2005: http://www.parlamento.pt/Legislacao/Paginas/ConstituicaoRepublicaPortuguesa.aspx

4 These socialist ideals are understood under the broad term 'social democracy'.

5 For Spain, whose history almost exactly replicates this timeline (both in terms of the duration of the dictatorial regime and the timing of its entrance to the EEC), Europe was also synonymous with modernization, with economic, scientific and technologic progress, and with the overall end of isolation. (Castañares and Juste 2002).

6 According to Sousa Santos, 'the convergence of policies of the European Union . . . and the structural and cohesion funds were imagined as a generous and efficacious political-economic device to allow Portugal a short circuit: to fulfill, in a few decades, her second Modernity to the full, something that had taken the developed European countries two centuries to accomplish. Thus, Portugal would coincide, with only a slight delay, with the rest of Europe in the tasks of transition toward the new modernity, transmodernity and postmodernity' (Sousa Santos 2009: 24).

7 Some authors, such as Pierre Rosanvallon (2008), have an even harsher interpretation of this type of data. For Rosanvallon, electoral abstentionism does not necessarily mean apathy or disinterest, but can point to its opposite, in other words, the expression of a position through the refusal to cooperate with current political and governmental decision makers. He goes on to explain a phenomenon he terms counter-democracy: 'By "counter-democracy" I mean not the opposite of democracy but rather a form of democracy that reinforces the usual electoral democracy as a kind of buttress, a democracy of indirect powers disseminated throughout society – in other words, a durable democracy of distrust, which complements the episodic democracy of the usual electoral-representative system' (Rosanvallon 2008: 8).

8 See http://www.planotecnologico.pt/InnerPage.aspx?idCat=31&idMasterCat=30&idLang=2&site=technological-plan (accessed 10 April 2013).

9 To learn more about the background of the 'Technological Shock', see http://www.choquetecnologico.com

10 Interview with the author, June 2012.

Chapter 2
Eye in the Sky

Is it possible that video surveillance (or even all modern technology used to monitor citizens), regardless of its practical effectiveness as an instrument in the action against criminal activity, represents an equalization of European cities? Given the compliance to a model of security adopted in most European cities, to what extent can public video surveillance be considered a symbol of modernization? Can the securitarian discourse itself also be a representation of that same modernization?

The British case undoubtedly constitutes an inevitable starting point for any study dealing with public video surveillance in Europe, both because the U.K. was a pioneer in this area, and because its long experience provides a rich source of information on the potential uses, benefits and weaknesses of surveillance devices. The developments that led to the expansion of its programme, for instance, may serve to contextualize the debate. To that end, I suggest that we go back to 1993, when two-year-old Jamie Bulger was abducted by two young boys (both were ten years old at the time) from the Strand Shopping Centre in Bootle, Merseyside, where he was with his mother. After maiming and killing him, the boys left the body by a railway track on the outskirts of the city, where it was found two days later. This event shocked the world, due not only to the violence and cruelty of the act itself, but also to the CCTV footage which made the reconstruction and actual viewing of every step leading to the fatal end possible. The internal CCTV equipment of this shopping mall recorded images of the child being taken by the hand by one of his kidnappers.

These images were later used as undisputable criminal evidence, in a demonstration of the efficiency and usefulness of cameras as technological devices at the service of law enforcement. Substituting the human eye and visual memory with a machine in this way was unprecedented. For the first time, CCTV – the abbreviation for closed circuit-television – demonstrated its importance and pertinence before a worldwide 'audience'. It showed how a crime could be reconstructed, the criminals identified, and furthermore, how the images could be viewed – as in fact they were – over and over. Surveillance cameras, unlike the human eye, offer the possibility of endless replay.

As Clive Norris and Garry Armstrong explain in their work *The Maximum Surveillance Society: The Rise of CCTV* (1999), before this incident, all political and governmental initiatives to implement large-scale video surveillance systems in public areas throughout the United Kingdom had been met with resistance, due to the inherent economic difficulties. These devices had so far only been adopted by small businesses worried about the loss of clientele to big multinationals operating in the recently emerging shopping malls, at a time when criminality was already a serious problem in major English cities. As Gavin Smith (2004: 378 and following) points out apropos of the economic recession in the United Kingdom during the 1980s and 1990s, traditional commerce in large cities was declining, given the shift of mass consumption to major shopping centres on the cities' suburbs, partly because of their increased and concentrated offer and partly because they were considered safer.

A few years after the tragic death of Jamie Bulger, as Norris and Armstrong show, the scenario had changed dramatically in British cities. Economic, political and popular opinion were coming together, pushing the fear of crime to the verge of a shift in paradigm: the security and safety of people and property would become the major concern; cities were open spaces of free circulation, and the public increased the tone of their demands urging that this investment be made as swiftly as possible. There was obviously a confluence of several interests at play here: public opinion and the political parties were in accordance, and business owners had nothing to object to – quite the contrary.

In no other European country can there be found anything that comes close to the high concentration levels of video surveillance devices, whether in public, semi-public or commercial spaces. These data are confirmed by the *Urban Eye* project (Hempel and Töpfer 2004). This is the most comprehensive multidisciplinary study on the presence of video surveillance systems in both public and semi-public spaces within Europe, and was conducted between 2001 and 2004 in seven European countries (Austria, Denmark, Germany, United Kingdom, Hungary, Norway and Spain). Its results reveal significant differences between countries in terms of the presence of video surveillance devices in open areas.

Specifically, there is a wide gap between the United Kingdom, which presented the highest numbers and widespread presence of such systems in most large cities, and the rest of the countries (McCahill and Norris 2003; Norris, McCahill and Wood 2004; Wood and Webster 2009). As stated in the report:

> In contrast to the similarities in the extent of CCTV in (semi)private spaces are the open streets where the deployment of surveillance cameras differs significantly

> across Europe. While it was estimated that around 40,000 cameras monitor public areas in more than 500 cities in Britain, less than 100 cameras monitor public areas in around 15 German cities, and no open street CCTV system is in operation in Denmark. (Hempel and Töpfer 2004: 5)

The study also emphasized the contrast between cities in the United Kingdom, France, Hungary, Italy and Monaco, where video surveillance systems comprise networks of over forty cameras, and the model adopted in countries such as Austria, Norway and Spain, where this type of device is limited to strategic locations, namely, major railway stations. The common feature shared by these countries is the political argument for the use of video surveillance as a means to deter criminal activity – as a crime-fighting instrument.

In Italy and Greece, two other European countries that have also been the subject of studies by sociologists working on CCTV use in open areas, it has been observed that this technology first emerged in the city of Milan, Italy, at the end of the 1990s. In a detailed study of this monitoring system, Chiara Fonio demonstrates how video surveillance on the one hand strongly influenced preconceived and discriminatory profiling of suspects[1] (in this case specifically of persons that appeared to be of Eastern Europe or North African origin), and on the other hand promoted voyeuristic behaviour on the part of camera operators, ultimately revealing their general lack of adequate training for this task. Reflections on the uses of CCTV contribute to our understanding of the reality-show effect that a technologically mediated observation of real events seemed to provoke on video surveillance operators.

> The 'border' between being an 'object of target' and an 'object of an a priori stigma' was slight and the labeling process seemed almost immediate: if one is a member of an ethnic minority is more likely to be watched because he or she is supposed foresee to represent 'a risk'. . . . Two social categories were also targeted on the basis of their appearances: young people, in particular those who were poorly dressed, and nice-looking women. (Fonio 2007a: 7)

The detached character of this visual relationship caused the monitoring of real-time footage of actual events to induce a state similar to watching a television show: in a sense, events appear to be a mere performance of successive episodes. Fonio emphasizes an aspect which deserves to be underlined: the incidence of unpredicted effects such as those witnessed when surveillance cameras are being monitored, including ethnic discrimination, suspect profiling and questionable physical scrutiny of persons (Fonio 2007b).

From among the extensive and valuable literature on these themes, I must point out, for the relevance it had on my own endeavours, the work of the Greek sociologist Minas Samatas, particularly his *Surveillance in Greece: From Anticommunist to Consumer Surveillance* (2004). Its historical study of surveillance practices in Greece became a reference due to its perspective on this country's political history (which has many parallels with Portuguese history) and its account of the transformations surveillance practices underwent throughout the decades, the multiple and many-layered effects on the everyday lives of Greek citizens, and their perception of these issues.

Following Gary Marx's proposition (Marx 2006), Samatas distinguishes between 'new' and 'old' methods of surveillance, and combines this analysis with an account of shifts in the common perception, especially between different generations, of these systems' interference with civil rights and freedoms. He therefore demonstrates that there can be significantly different outlooks on surveillance practices, and that these can be attributed both to the specific features of a given country and to its historical, economic and social circumstances. This approach is also followed by authors such as Calenda and Fonio who explain that 'while it is true that most scientific studies belong to north-American and Anglo-Saxon production, focusing above all on an institutional context hardly comparable to the Italian one, southern European countries such as Greece, Spain and Portugal, share a post-dictatorial legacy that offers tremendous potential for comparison' (Calenda and Fonio 2009: 119).

The Greek case is illustrative of how video surveillance systems, conceived as technological mechanisms to reinforce security, do not always produce favourable or even predicted results. Referring specifically to video surveillance in public places, Samatas describes how the city of Athens, when hosting the 2004 Olympic Games, was equipped with over 1 200 sophisticated video surveillance devices. This was an operation in which the Greek government suffered considerable international pressure to guarantee security conditions with a view to the safety of the Games' proceedings, particularly from the U.S.A. which was at the time fully engaged in the 'war on terrorism'. The result of such costly investment is that several years later, the majority of the cameras were inoperative, either due to a shortage of qualified personnel to run them, or because neither the state nor private companies were able to sustain them financially. The compromise solution found for this standstill was to sell the cameras to private entities, to be installed along the major roads which every year register high death rates due to traffic accidents.

According to Samatas, in this case, the decision, while valid, failed to act directly on the causes behind this phenomenon, in this particular case the roads' bad state of conservation, deficient lighting and poor access. Samatas also argues that, on the other hand, all the investment in security devices aimed at the very specific and temporary event of the Olympic Games could have been more advantageously used to improve other more urgent areas such as better infrastructures for the education and health sectors. Finally, Samatas questions this large investment in surveillance cameras, pointing out the strong opposition that it raised among civil society because 'the memory of authoritarian and repressive surveillance used in the past for thought control, and the fact that police CCTV cameras represent the central state and the security forces, now cause fear and mistrust' (Samatas 2008: 360).

Samatas' description provides a paradigmatic instance of a case where the implementation of video surveillance was essentially a political phenomenon, and, as such, very complex and controversial. As will be seen further on, the Portuguese case exhibits similar characteristics, clearly demonstrating that we are dealing with phenomena which, far from forming a consensus or mobilizing shared interests, reveal the conflicting relationship that has developed among the different parties involved in various types of power struggle.

The Eye behind the Eye

I wish to add to this brief bibliographical review a reference in connection to a pertinent theoretical discussion around the video surveillance device. Notwithstanding the abundance of empirical case studies and theoretical works on surveillance practices in contemporary society, the works of Michel Foucault remain to this day one of the most important benchmarks in this field. Whether admired or criticized, he is still probably the author most often cited and discussed by scholars in surveillance studies, as Zureik, Lyon and Abu-Laban have accurately observed: 'Surveillance Studies has developed apace, especially in the countries of the global North. Theoretical development, following Foucault in the first place but then based on the critique of Foucault or even the rediscovery of older social and political theorists, has occurred in tandem with the expansion of empirical studies' (Zureik, Lyon and Abu-Laban 2011: 20).

The reactions to Foucault's work in this field over the last decades are particularly to his 1975 book *Surveiller et Punir*, published in English under the title *Discipline & Punish* (1995) – even though the French term *surveiller* is the literal equivalent of the English term 'survey', as pointed out by Klauser (2009b).[2] One of the reasons why *Discipline and Punish* stands out as an obligatory reference for any study on video surveillance – despite this philosopher's other works also dealing with issues of control, governance and power, such as *Securité, territoire, population* (2004) or *Society Must Be Defended* (2003) – is his analysis, in this specific work, of disciplinary techniques, and above all his interpretation of a device dating back to the eighteenth century: the panopticon.

The panopticon was an architectural model developed by the British philosopher Jeremy Bentham around 1785, conceived essentially as a device to improve the surveillance capability of prisons, but which could be used in any kind of facility where vast agglomerations of people gathered, such as schools, hospitals or factories. It consisted of a circular structure built around a central tower from which it was possible to carry out constant and unidirectional monitoring surveillance over the whole building. Two centuries later, Foucault describes this model as a power device based on the ability to discipline, not through physical punishment, but merely through the subjects' complete exposure to another who remains invisible and unknown.

As such, surveillance alone gains disciplinary force insofar as it determines a behavioural change in subjects without the need for direct interaction. In the author's words:

> Hence the major effect of the Panoptic: to induce in the inmate a state of conscious and permanent visibility that assures the automatic functioning of power. So to arrange things that the surveillance is permanent in its effects, even if it is discontinuous in its action; that the perfection of power should tend to render its actual exercise unnecessary. . . . To achieve this, it is at once too much and too little that the prisoner should be constantly observed by an inspector; too little, for what matters is that he knows himself to be observed; too much, because he has no need in fact of being so. In view of this, Bentham laid down the principle that power should be visible and unverifiable. (Foucault 1995: 200)

These features led several authors to discuss the role the panopticon represents as the direct ancestor of today's surveillance cameras, whose power to dissuade crime or other activities considered deviant and thus to normalize the behaviour of anyone who believes to be under their gaze rests on the devices' mere presence.[3] The panopticon as a metaphor of omnipresent power and the idea of cameras as its materialization have gathered over the last years an equal share of enthusiasts and sceptics, divided in their answer to some key questions. Are surveillance cameras a modern version of the panopticon? Are these devices the material analogue of Foucault's theories? If so, to what extent is this author's analysis of their potential uses accurate, namely that their disciplinary effect derives from the act of constant monitoring?

Although Foucault ended up virtually abandoning a perspective centred on disciplinary devices to pursue a more theoretical line focused on security, governance and bio-power (Bigo 2008; Dillon and Neal 2008; Giddens 1985; Lyon 2006; Razac 2008), the way in which his theory of the panopticon equates visibility and power remains extremely insightful.

Following the path laid by Foucault, Gilles Deleuze, in his much cited article 'Post-script on the Societies of Control' (1992), went even further with the analysis of discipline, by claiming that it has gradually been replaced by the 'desire' to control. This idea constitutes a shift in paradigm insofar as the goal of discipline would cease to be the rehabilitation of transgressors by means of an 'orthopaedic discipline', as Cunha describes it (Cunha 2008, 2002), to become the prevention or anticipation of wrongful activity or, failing that, the punishment of perpetrators (Garland 2002). Nikolas Rose also assumes this theoretical turn when he states that '[t]he new techniques target offenders as an aggregate; they do not aim to rehabilitate, reintegrate, retrain, provide employment for particular individuals. They seek only to reduce rates of crime and risks posed by groups – such as the urban poor or underclass – by whatever means are appropriate to the risk they prevent' (Rose 1999: 236).

The discussion here has a double implication. In a certain sense, Foucault sets out the grounds for the idea that the act of seeing, or in this case surveying, symbolizes power that is instrumentalized to discipline, control or monitor individuals; I stress that I am strictly applying this definition to the idea of the panopticon, and not the manifold implications of broader issues related to vision and observation. At the same time, surveillance cameras are devices built precisely for that purpose, which the urban dweller comes across in the various situations of daily life, from commercial areas to working environments, in schools, hospitals, public transportation, etc. There is no question that there are resemblances between the panopticon and the surveillance camera, one of which is what gives the latter its enormous potential: the camera, unlike human mechanisms, is able to capture and record images constantly and with a precision and detail that is beyond what any human can do. It allows endless rewinding and replay, powerful zooming and a whole set of technical possibilities, all the while maintaining the observer hidden from the subject under surveillance.

The trope of vision is useful here insofar as the surveillance activity by means of cameras prefigures acts of control, observation and constant monitoring which will probably carry consequences for the subjects of that surveillance, whether they are individuals or territories (Brighenti 2010). It functions as a power mechanism in the sense that it is exerted unilaterally. Those under surveillance, even when aware of it, are in most situations powerless to return the gaze, which would in any case reveal nothing. Picture, for instance, a situation in which someone walking down a street, and happening to notice that he is being monitored by a surveillance camera, decides, as a form of reaction, to stop in front of it, staring back and trying to engage visually with whoever is supposedly watching behind the camera. This subject may even successfully draw the supposed camera operator's attention and convey his message of 'Hey, I know you are there watching me'. The obvious difference is that, while the operator's capacity to see the person on the other side enables him/her to effectively establish some sort of identification, and thus s/he can be said to gain direct contact, the person in the street can merely display a reaction while seeing nothing more than the lens of a machine (Koskela 2003).

Video Surveillance in Portugal: Law No. 1/2005

As described earlier, when Jamie Bulger went missing from the shopping mall it was the images captured by this commercial facility's CCTV system that made it possible to identify his abductors and bring them to justice to be punished for their crime. The impact of this episode on the public opinion – amply covered by the media – triggered the widespread installation of video surveillance in most British cities, or in the absence of cameras, the emergence of neighbourhood patrols constituted by concerned citizens who watched over their neighbourhoods – the well-known Neighbourhood Watch programmes (Abrahams 1998; see also Pratten and Sen 2007).

Over twenty years later, also in the United Kingdom, the multiple video surveillance devices spread across London were instrumental in the identification of those responsible for the terrorist bombings in the city's public transportation system (underground and bus) that injured over seven hundred people, killing around fifty. Events like this effectively motivate and explain the occurrence of a sometimes radical shift regarding the issue of surveillance, both in terms of individual behaviours and in terms of shaping public policies. Consider the impact and consequences of U.S. policies in the post-9/11 context: the twin towers attack was an event that was indeed responsible for a new paradigm of world security. These examples to a certain extent serve to illustrate the process, reasoning and aim of the installation of surveillance devices in general, and in the case of Portugal, of video surveillance dissemination as a response to emerging or unforeseen events in the daily lives of the population, especially in urban areas. Nevertheless, the Portuguese case displays certain nuances that are worth noting regarding the alleged purpose and goals of video surveillance implementation and even the motives behind this initiative.

In Portugal, video surveillance in public spaces was initially conceived as an extraordinary measure, since its use – with regard to the invasion of privacy and its impact on the rights and freedoms that generally characterize modern democratic societies – could be justifiable only under exceptional circumstances. As Agamben (2005) explains,[4] exceptional phenomena call for exceptional measures. The first time that the use of video surveillance cameras in public spaces was contemplated was on the occasion of the 2004 European Football Championship ('Euro 2004') hosted by Portugal. The event provided just such an exceptional circumstance to open the way to legal measures that superseded current legislation and allowed its temporary suspension.[5]

The implementation of summary judgements and extraditions would hence become possible in this context. Specifically, for the duration of the competition, police forces were authorized to make an uninterrupted sound and image recording of the stadiums' premises as a means to ensure the smooth running of the event and public safety. The most significant aspect of this episode – this epiphenomenon – is that, albeit exceptionally, individual rights were effectively suppressed, allegedly as a means to protect a greater good: collective security. Although Euro 2004 ended in July of that year, the law enabling the project of video surveillance installation in public spaces to go ahead was not passed until January 2005, less than six months later. As in Greece during the 2004 Olympic Games, where millions of euros were invested to turn the city into a kind of fortress, shielded against all sorts of potential threats (Samatas 2004), the football championship in Portugal served as a trial case to assess the benefits of these devices and their potential use on a wider scale. Whereas in Portugal video surveillance previously had been restricted to confined spaces (e.g., commercial areas, petrol stations or banks) and was managed by private security companies, Law No.

1/2005[6] authorized the implementation of this technology in public spaces. It also gave the Public Security Police and the National Republican Guard the authority to monitor and register the images collected in these areas. Video surveillance is interpreted, within the spirit of Law No. 1/2005, as an auxiliary mechanism intended to help fight crime and improve police efficiency and performance. The promotion of this measure rests on the premise that the use of video surveillance devices will have a significant impact by deterring and preventing criminality and that it will be an extremely reliable instrument for the identification of criminals in post-crime investigations.

The cameras in question are equipped with a zooming power that covers up to a range of five hundred metres, a rotation of between 180° and 360°, and a built-in filter to block out images that capture doorways and windows of private residences (Figure 2.1). Recorded footage is monitored and viewed exclusively by police force operators and destroyed within a period of thirty days after it is taken, except when it registers a criminal act, in which case it is preserved and may be used by court order as potential means of evidence. Monitoring rooms are located inside police force premises, and as a rule they are set up as closely as possible to the areas where the cameras are installed. Operators receive specific training in video surveillance monitoring. A register is kept of the content and time of the material viewed by operators, and all monitoring activity is supervised by a superior officer as a measure to prevent the occurrence of irregularities and violations (Figure 2.2).

Let us now see precisely how the permission to use public video surveillance is actually defined by Law No. 1/2005. The terms of the National Video Surveillance Programme will also be analysed, so that later a more accurate assessment can be made of its application during the period between 2005 and 2010, in what might be called this phenomenon's initial stage in Portugal. As the law conceived it, video surveillance in public areas of common use was designed to ensure: 'a) Protection of public property, such as public buildings and their premises, including access roads and entry or exit routes; b) Protection of facilities that serve national security; c) Protection of people and goods (public and private), ensuring their security and preventing crime in areas that present a reasonable risk for its occurrence'.[7]

This law also decrees that requests for installing cameras can be made only by the chief officer of the security force responsible for the area in question or by the governor of that same area in collaboration with security forces. After approval by the Ministry of Internal Affairs, the request goes on to the Data Protection Authority, which determines whether the project is in accordance with the standards of personal data protection, and then makes a final ruling that has a legally binding force. The law also stipulates that authorizations, once granted, have a one-year maximum duration and may be suspended, revoked or eventually renewed at any time 'upon confirmation that the reasons invoked for its concession persist'.[8]

Which guidelines should authorizations to use video surveillance cameras comply with in this particular context – that is, public space? According to

Fig. 2.1: *A Dome Camera in the Oporto Historic District*

the legislation, the guidelines must follow a principle of proportionality, defined by the following conditions: (1) the usage is permissible 'whenever this type of equipment proves the best-suited to ensure security and public order and to prevent crimes, and always taking into account specific features of the prospective area under surveillance'; (2) 'the probability and extent of interference with personal rights' must always be taken into account; and (3)

Fig. 2.2: *The Oporto Control Room of the Public Security Police*

the use of concealed cameras is prohibited, and the installation of 'video surveillance cameras must always imply the existence of a real threat to security and public order'.[9]

The National Video Surveillance Programme

Two years after Law No. 1/2005 was approved, the National Video Surveillance Programme was first made public in 2007 by Rui Pereira, the minister of internal affairs, as part of the Safe Portugal Strategy. At the time, Pereira stated that, among other measures intended to update public security forces technologically – encompassing the recruitment and training of manpower and the acquisition of more sophisticated firepower – the installation of video surveillance in public areas all over the country was also being considered. The proposal was presented by the minister of internal affairs as follows:

> The use by police forces and services of video surveillance and other electronic means is strictly destined to work as an asset in the correct performance of their mission to serve the community and improve collective security, with full respect for the rights implied by the protection of private life.
>
> This Program contemplates:

– Legal authority for security forces to use video surveillance devices, mobile or fixed, in various missions, especially for the prevention and prosecution of traffic violations.

– The access of Security Forces to data collected by Portuguese Road Services PLC (Estradas de Portugal) and by its concessionaries, through their systems for detecting accidents or violations in real time.

– The constant monitoring in situation rooms by the managing entities of road traffic on major road axes.

– The use for specific purposes of electronic systems to control incidents and accidents, by the managing bodies of national roads.

– Security forces may access and use electronic surveillance at a municipal level.

– The possibility of placing means of video surveillance for ensuring the safety of cab drivers.

– The legal authorization to use video surveillance to comply with legal norms, namely, criminal detection of stolen vehicles, and fake registrations for proofing purposes.

– Automatic number plate recognition of vehicles in circulation that do not meet the legal requirements.[10]

As is evident from this statement, the National Video Surveillance Programme, a project whose resonance would apparently reflect the state's strong resolve in this matter, after all amounted to a thin and vague document, which can at best be described as a short list of guidelines for the project's execution. It is devoted essentially to matters regarding road traffic monitoring. In this sense, when we specifically consider the effectiveness of Law No. 1/2005, and its articulation through the National Video Surveillance Programme, we realize that the proposed goals of this law are to some extent thwarted by the very features that underlie the state apparatus itself.

To begin with, whereas, on the one hand, the name given to this government initiative points to the installation of video surveillance cameras throughout the whole national territory, to be distributed according to requests made by the different municipalities, on the other hand, the Data Protection Authority does not have the same interpretation or understanding of the law in question. In fact, as will be made explicit, the idea of introducing CCTV systems in urban areas is frequently politically motivated, and the largely subjective arguments with which it is supported – such as the desire to fight perceptions of insecurity – may be described as essentially populist. This type of subjective reasoning, however, clearly does not meet the criteria set by the law that, among other things, demand the demonstration of objective risks to security and safety by means of crime rate assessment – data that are, in turn, used as the basis for determining the inadequacy of said initiatives.

Let us now analyse a few instances of this law's application, in some cases more successfully than in others, not so much to judge the efficiency of surveillance cameras as a security system, but to introduce the conflict underlying the relationship between the Ministry of Internal Affairs, the police, and

Table 2.1: *Decisions on Video Surveillance Projects Submitted between 2005 and 2010*

Year	City/Area	Number of Cameras	DPA Ruling	Final Decision
2007	Oporto	15	60/2007	Approved (21:00–07:00 h)
2008	Portimão (Praia da Rocha)	20	21/2008	Approved (8 cameras only for the car park)
	Batalha	1	27/2008	Rejected
	Fátima (Sanctuary)	8	36/2008	Approved (06:00–02:00 h)
	Coimbra	32	47/2008	Approved (12 cameras 20:00–08:00 h)
2009	Bragança (Industrial Area)	3	04/2009	Rejected
	Estarreja	3	29/2009	Rejected
	Baixa Pombalina (Lisbon)	32	62/2009	Rejected
	Bairro Alto	27	68/2009	Approved (6 months, 22:00–07:00 h)
2010	Amadora	113	48/2010	Rejected

the Data Protection Authority, which will be developed in the following chapter. The next few pages will offer an account of the attempts to install video surveillance devices in public spaces at different locations throughout the country, and with different degrees of success. What will clearly transpire from this account is the tribulation that marks the whole process and its vicissitudes, which will be discussed in greater detail at a later stage.

Ten requests for video surveillance in public areas were submitted between 2005 and 2010, out of which five were granted (Table 2.1). By the end of 2010, three of these video surveillance systems were operational – in Oporto and Coimbra's historic districts and in Fátima's Sanctuary – while in Lisbon's Bairro Alto district, installation was underway. The projects that were rejected included those in the cities of Estarreja, Bragança and Amadora, and also Lisbon's Downtown district, the Baixa Pombalina. As this data clearly indicates, even though the promulgation of the law that allowed installation of surveillance cameras in public areas dated back to the beginning of 2005, the first time such a system was effectively authorized to operate was in 2007.[11]

It should also be noted that 2008 and 2009 were years that registered several processes ongoing simultaneously, suggesting greater interest in this type of measure by local leaders and the police during this period. In 2010, the only project to be submitted was rejected in the end.

If an analysis were made based merely on the quantitative data, this trend could easily find a probable cause, especially following the obvious rationale of the end served by video surveillance, its relation to crime rates and feelings of insecurity among the population. It would have to be concluded that the decrease in these indicators had logically led to a diminished interest by authorities in using surveillance cameras throughout Portuguese cities. The real explanation is actually more mundane, but that is why I chose to present these numbers straight away, before showing how easily they can be misread.

There was in fact a growing disinterest in these projects, but the reason is not so much bound up with criminality – which is indeed low – or a perceived lack of necessity, but instead with the obstacles met by these requests in their last stage; that is, obtaining the Data Protection Authority's final seal of approval. Although it could be argued that there is a balance between the number of positive and negative rulings, the emphasis falls strongly upon on the latter; first because even positive rulings were only partially authorized while negative ones were absolute, and second because in most cases, candidates were baffled by the Data Protection Authority's decisions, which became generally perceived as unfathomable, thus creating an atmosphere of scepticism.

The following section will consider in greater detail specific features that may prove decisive in granting or rejecting the use of video surveillance in areas otherwise presenting similar characteristics. The Oporto Historic District (Zona Histórica do Porto, also known as Ribeira do Porto) and Lisbon's downtown area (Baixa Pombalina) – both being important touristic sites of historical interest and patrimonial value – are clear examples of this, allowing us to identify the factors that must be pondered in each decision and to establish the relevance attributed to video surveillance in public areas as an effective device for the protection of persons and goods in each case. These cases will now be considered carefully, and the possible causes behind the divergent outcomes will be discussed.

Given that the areas under consideration are geographically scattered and the objectives clamed for each case diverse, examples will be used that may serve to illustrate the criteria and distinctions used to support the final decisions. Besides an ethnographic description that provides an accurate timeline and a thorough depiction of the different interlocutors' involvement and positions – which would not be possible by mere analysing official documents and rulings or even through isolated interviews – some relevant issues for analysis will be discussed. These include the importance of political and media discourses and of public opinion regarding security-related matters; the prevailing economic situation; and, lastly, the delicate matter of legal adequacy regarding the use of video surveillance, especially considering its interference with basic rights and freedoms of citizens. These general themes roughly summarize the issues that will be an object of extensive discussion over the next few chapters. But before that, this chapter will focus on giving a factual account of the developments that occurred over the five years under analysis, as a way to give further emphasis to the historical context already considered in the previous chapters in terms of the country's political evolution, including the successive strategies of modernization in both their political and material implications, particularly regarding the use of technology in crucial areas of social life.

Video Surveillance in the Oporto Historic District

The first case that Law No. 1/2005 was applied to was in the Oporto Historic District. This area is approximately 1 250 m^2, stretching along one of Douro River's banks, it occupies a significant part of the city's historic centre. Still maintaining its medieval urban design and having played an important role in the city's shipping industry during the nineteenth and twentieth centuries, it can be described as a tight network of narrow streets and poorly lit alleyways (Fig. 2.3 and Fig. 2.4). Over the last two decades it has become above all a touristic spot, displaying an intense activity of all related businesses to meet a demand that flows in throughout the year, both for daytime and night-time activities, especially during the summer season when its pleasant riverside esplanades become crowded. Despite being highly sought out in the daytime, this district is chiefly famous for its nightlife, attracting mostly a young population, both local and foreign.

Between the end of 2005 and the beginning of 2006, António Fonseca, the president of the Commercial Association for the Oporto Historic District (Associação de Bares da Zona Histórica do Porto), publicly claimed that, considering the area's status as a UNESCO World Heritage Site and a touristic point of interest, the city council should ensure the district's security by installing a CCTV system to improve police force performance. Initially, the mayor's office was not wholly enthusiastic about the proposal, characterizing the video surveillance as unnecessary and invasive of personal privacy and also taking into account the substantial financial investment that would be involved. An open and public conflict had nevertheless been started in which the association's side relied on the citizens' feelings of insecurity and the overall perception that governing authorities were apparently doing nothing to improve the situation, abandoning businesses to their own devices and forcing them to hire private security and internal CCTV systems.

During 2007, and as a result of the constant pressure from this association's members, the city council began drafting an official request to install a video surveillance system for that specific area of the city, to be financed by the Commercial Association itself. This idea was well received by Oporto's Public Security Police Department and was immediately accepted. The police official responsible in Oporto, who drafted and substantiated the project based on local crime rates, later revealed to me that he was truly enthusiastic about it. If the receptiveness shown by both police forces and the city council is to be fully understood, the underlying regional rivalry (particularly with the capital city, Lisbon) that to a certain extent motivated this process must be considered. In fact, the parties whom I interviewed freely admitted this to me. For a city admittedly obsessed by its eternal second-place national

Fig. 2.3: *An Alley in the Oporto Historic District*

ranking in terms of economic and demographic significance,[12] the chance to pioneer the implementation of this 'modern' security device makes this element of rivalry worth considering in this particular case.

In rough terms, the city council's proposal called for a 24/7 video surveillance recording of both sound and image covering the area in question. The request argued that this particular area 'had over the last years been suffering from an increasing state of insecurity [and] that it was a stopping point for

Fig. 2.4: *Overview of the Oporto Historic District and Douro River*

drug addicts'.[13] It also described the area's urban geography as predominantly consisting of 'narrow, poorly lit alleys'. Regarding crime rates, it claimed that Oporto – not only this area but the city as a whole – had '30 per cent higher rates than other municipalities'. Criminality was described as mainly consisting of 'auto theft, petty theft (purse-snatching and pickpocketing), burglary, and damage to cultural patrimony'.

Furthermore, part of the reasoning used to justify the implementation of video surveillance in Oporto was the fact that it was 'referenced in several national and international touristic guides as a mandatory site to visit'.[14] The proposal described the specific locations for the cameras, as well as the system's technical characteristics, and appointed the Public Security Police as the entity responsible for its monitoring. In February 2007, the proposal was sent to the Ministry of Internal Affairs and evaluated. The resulting report stated that this proposal was 'not only of significant interest for local security but also undoubtedly urgent'.[15] The whole project was subsequently remitted to the Data Protection Authority for a final decision. Several months later, in December 2007, this institution finally granted what, in view of the initial request, may be characterized as a partial authorization. It agreed to video surveillance monitoring but denied the use of sound recording; the ruling also limited the surveillance to the night-time hours (from 9:00 pm to 7:00 am). Furthermore, the implementation of video surveillance was subject to a trial period that was legally determined to be a maximum of one year.

It was in January 2009 that I first requested to visit the site. As the news media had announced that there were video surveillance cameras in Oporto Historic District, I was surprised when, at the office of the Public Security Police of Oporto, I was told that the cameras were indeed there but were not operating. My visit was nonetheless welcomed, I was assured, and assistance would be freely granted. The superintendent in charge explained to me that the cameras had been at their spots for over a year, but that, due to financial and logistical reasons, it had not yet been possible to operate them. As mentioned earlier, the city council had supported the request for a video surveillance system politically but not financially, and although the Commercial Association had agreed to finance the system, it turned out that it did not possess the means to fund the wireless or cable connection that was necessary to ensure proper monitoring conditions, namely, regarding image definition and real-time transmission to police stations. Police officers jokingly said: 'Well, at least the cameras were not vandalized. Now that one year has passed [the stipulated trial period] we can safely report to the Data Protection Authority that it has been a success!'

Several months later, in September 2009, when apparently all problems had been resolved, the inauguration of a video surveillance system in Oporto was publicly announced. But one significant detail had been overlooked and was not yet settled – the matter of who was going to pay the electricity bill for the energy needed to run the cameras. Thus, it was not until November 2009 that the system effectively began functioning, with the Oporto Historic District being monitored and recorded by Public Security Police, in fact, two years after its approval by the Data Protection Authority.

I returned to that city in January 2010 to continue my conversations with the Public Security Police and the mayor's office and to proceed with my planned fieldwork. At this stage, I was somewhat disappointed to learn that I would not be given access to the rooms where CCTV operators monitored the images in real time, given that this would require a special authorization from the Data Protection Authority. Still, I was given the opportunity by members of the police to view some prerecorded footage, shown to me in the spirit of emphasizing the kind of problems they were facing at the time, which were mostly technical difficulties linked with the quality of the system's image. To begin with, obtaining a clear view of the areas under surveillance was made hard due to deficient or even inadequate surrounding lighting, as for instance in situations where cameras were blinded by light reflecting from public light sources directly pointed at them. In addition, inadequate optic fibre connections resulted in delayed transmission, thus making real-time monitoring and swift police action impossible. Overall, image quality was so poor that it became impossible to respond effectively to any occurrence. Furthermore, images transmitted by cameras are displayed on screens that are like small windows through which one can glimpse only a small fragment of a given area. Even in limited spaces, however well equipped with cameras and monitoring agents (and this was not the case here), for the police officers, it was impossible to keep a constant observation on all the screens and details, especially when there was more than one action occurring at the same time.

In terms of how the security forces viewed the efficiency and necessity of this system, opinions were divided. Statistics showed that the type of crime in this specific area was typically limited to minor offences (such as pickpocketing and public disorderliness usually connected to alcohol abuse), which did not justify serious police intervention. Moreover, police forces acknowledged that the insignificant numbers of incidents of public disorder registered in 2008 – the year in which the installation of the video surveillance system began – were confirmed by similar data in 2009, except for an extremely important nuance: the flow of people had drastically diminished, especially in the off-season after the summer months. These low numbers were thus apparently in no way connected to the effect of video surveillance, as had been hitherto foreseen.

What was described by the police as the desertion of Oporto's Historic District, traditionally known for its active nightlife, was instead due to the fact that the young and itinerant people who had traditionally occupied this locale had found new places of recreation in other parts of the city. Therefore, the 'increased' feeling of security that had been expected to result from the newly installed video surveillance system was rendered virtually irrelevant. The inactivity captured by the cameras was due mostly to the emergence of new bars and discos in other parts of town, rather than to any contribution made by video surveillance to the area's (now reinforced) security. As some Oporto police officers in charge of monitoring the cameras described it: 'We see nothing because there is nothing to see'. Thus, while some officers claimed that it was a ridiculous waste of time and effort to be watching monitors where nothing happened for hours on end, when they could be performing other more useful activities, others somewhat optimistically concluded that this information was in itself useful: 'Knowledge that nothing is happening is still better than no knowledge at all'.[16]

The discourse of the intervening parties in this particular case focuses on a proactive approach and on crime deterrence, rather than on forms of reactive intervention. As stated by Lucia Zedner: 'Conventional reliance on reactive post-hoc measures is being overlaid by a trend towards new preventive measures that seek to anticipate and forestall harms long before they occur and to maximize security pro-actively' (Zedner 2009: 73–74; Zedner 2007). In accordance with this perspective, members of security forces claim to favour types of knowledge that allow them to anticipate crime, instead of being passive subjects, acting only after such crimes are officially reported. In this case, it is easy to recognize that the pressure exerted by businesses played a decisive role in helping the process of public video surveillance to be developed and put in place. Nevertheless, the encouragement to install surveillance cameras was not backed up by any previous diagnosis based on actual crime rates; instead, it resulted from the convergence of interests expressed by political and economic parties.

The following case, which deals with video surveillance in the Lisbon downtown area, presents a process with similar characteristics, marked by a focus on concerns with the security and protection of persons (both resident and transitory populations) and patrimonial assets. However, it will become clear that the results obtained by this project were very different and present striking contrasts with the previous one.

Video Surveillance in the Baixa Pombalina

In January 2009, I met for the first time with the president of São Nicolau parish, who had submitted the proposal for video surveillance in the Baixa Pombalina, which is the Lisbon downtown area. This part of the city, which extends over approximately 23.5 hectares, is a major national and international tourist attraction, known for its architectural patrimony dating back to the eighteenth century and characterized by its dense commercial activity.

The initiative to request video surveillance for this area came from the president of the parish himself and was actually included as part of his electoral campaign back in 2005. Ever since the law permitting public video surveillance was passed, the parish president has made every effort to implement this system. However, despite counting on the direct support of the Ministry of Internal Affairs for his endeavours, he told me that he had ignored the existence of the National Video Surveillance Programme and its goals. His proposal had not been coordinated with the other entities involved in this process, such as the Data Protection Authority and the Public Security Police. Instead, it constituted an isolated and ad hoc action. The president described the arguments that supported his request as follows:

> This is one of Lisbon's tourist areas *par excellence*: over two hundred thousand people cross it every day. It includes many businesses and an architectural patrimony that must be preserved and protected by a strengthened security system. The cameras will not replace police officers, but they will help to meet the needs we have in this area.
>
> There is a general feeling of insecurity; shop owners complain that business is dwindling because customers are fleeing to other areas where they feel safer. And obviously the police force cannot be everywhere. It doesn't have the means to monitor and act with the efficiency that a video surveillance system can provide.

In his somehow unofficial proposal to the Ministry of Internal Affairs, and following a report made by foreign specialists, he estimated the need for thirty-two cameras to monitor the whole perimeter of the parish. He also contemplated the system's expansion, at a later stage, to other adjacent areas in order to ensure the security and safety of the whole area in question. The Baixa Pombalina consists of a series of parallel avenues (one of which has been converted to a pedestrian street) crossed perpendicularly by a set of narrower streets that are very poorly lit at night.

The parish president's confidence in a successful outcome for his request was based – in theory, not unreasonably – on the fact that this was one of the Portuguese capital's most popular touristic and commercial sites: 'No one [Portuguese or foreigner] leaves Lisbon without having taken a walk through the Baixa Pombalina!' he asserted. In his opinion, video surveillance should be considered essential in this area – similarly to other European cities – as an indication of the modernity of the country itself.[17]

After receiving and studying the project, the Public Security Police wrote in their report that 'this is socially a highly critical area, with an aged resident population that is particularly sensitive to issues of security, and where criminal occurrences generate great social alarm and reflect acutely on the citizens' everyday lives'. The agency also added that the sharp distinction between daytime and night-time populations on the streets greatly contributed towards a feeling of insecurity 'despite the constant efforts of Public Security Police officers'.[18]

Following several exchanges between the entities involved – the Ministry of Internal Affairs, the Public Security Police and the Data Protection Authority – the decision of the Authority was publicly announced in November 2009. The ruling was completely negative, finding that the request 'does not clearly demonstrate its conformity to the law', especially considering that the data regarding crime rates were negligible in proportion to the high numbers of people who cross the area daily, having in fact diminished over the previous two years. The argument that video surveillance would help to minimize the feeling of insecurity was equally dismissed, based on the claim that '[feelings of insecurity] do not always correspond to actual insecurity and can generally be felt by people who have not suffered criminal occurrences, which must in any case always be gauged in terms of objective data regarding concrete crime rates'.[19]

The president of São Nicolau parish told me that this was 'good news for pickpockets' who would continue to go unpunished. There was a second request for an even larger area and almost double the number of cameras. In January 2011, this proposal was rejected as well with the same argument that the perimeter/number of cameras was disproportionate to the crime rates in that area.

Conclusions and Comparisons

After considering these two cases of the Oporto Historic District and Lisbon's Baixa Pombalina, some conclusions and some contrasts can be drawn. Apparently, both urban areas are similar, being popular locations in major Portuguese cities with high numbers of tourist visitors (internal and foreign) and intense commercial activity. Given these similarities, how can it be explained that one of the projects was granted (even if limited in terms of coverage, since it would only function during the night-time and not around the clock as initially requested), while the other was rejected out of hand? A careful analysis of both requests immediately reveals certain elements that set them apart in relation to three major issues: official crime rates, number of people to be monitored (including resident populations and visitors), and the extension of the area in question.

While the request for the Oporto Historic District involved an area that may be considered both relatively small and clearly demarcated and was supported by crime-related data and documentation illustrating the relevance of the motives, the project for the Baixa Pombalina did not present an equally convincing case, especially regarding its feasibility. Firstly, the official crime rates that were presented for the Baixa Pombalina were insignificant in relation to the extremely high numbers of people allegedly frequenting this area. Moreover, the proposal did not convincingly demonstrate that video surveillance would provide an effective solution for the predominant type of crime in that area, described in the request as pickpocketing, purse-snatching and car burglary.

Another line of argument that had the opposite effect of what was intended – since it was not contemplated in the law – regarded the resident population's feelings of insecurity, as well as the parish's intention to renovate this part of town, making it more attractive for new residents. The fact is that any kind of feeling, whether of security or insecurity, is by definition a subjective notion. It may be grounded in facts – crime rate increase or having been the victim of a crime – but it may also result from a phenomenon of contagion spread by people, who in turn may or may not have experienced some kind of crime, or conveyed through mass media. In addition, urban rehabilitation belongs to the sphere of urban planning and is not considered by law to be a valid argument for authorizing video surveillance in public areas.

As will be discussed in the next chapter, according to the Data Protection Authority, this argument would only be valid in cases of exceptional criminal activity since the use video surveillance in open areas is itself bound by the exceptional nature of this measure. In other words, opposed to what municipalities and police forces apparently believed, judging from the projects they submitted, the lack of proportion between the population's high density and the low crime rates in the area proved to be a disadvantage rather than an advantage. Having a large population and an extensive area to protect do not necessarily assure the granting of video surveillance. On the contrary, the authorization of a given project is determined by the 'existence of objective risks' and clearly defined perimeters.

The Protection of Thousands

To counterbalance the example of Baixa Pombalina, and still focusing on requests for video surveillance that mention the protection of 'thousands of people', we will now consider the basis for authorizations awarded in two cases that are as different from each other as the kind of activity carried out in each one: the Sanctuary at Fátima, with a system operating since 2009 (renewed in 2010), and Lisbon's Bairro Alto, whose permit was issued at the end of 2009, though only for an initial six-month trial period.

Both these cases provide relevant clues toward clarifying the motivations that underlie the Data Protection Authority's rulings, which at the first glance would seem to be purely casuistic. How can it be explained that Oporto Historic District has video surveillance while the Baixa Pombalina does not, considering that both are notable areas in their cities? What is the reason that in some cases the high population factor results in rejection while in others it counts favourably?

The concentration of hundreds of thousands of persons over definite periods of time has been considered to justify the need for additional security mechanisms in both the Sanctuary at Fátima and the Bairro Alto district of Lisbon. The project for the Sanctuary at Fátima, whose eight cameras were fully working by the beginning of 2009, was actually the first in Portugal to have an operational system. The Sanctuary at Fátima is one of the most important shrines in the world dedicated to the Virgin Mary (where it is believed the Virgin made several apparitions in 1917). This sanctuary welcomes millions of pilgrims and tourists from all over the world who come to visit the basilica every year, especially on 13 May and 13 October, the dates of the first and last apparitions respectively.

The initial request to install video surveillance in this area did not come from either the police forces or the city council, but from the entities responsible for the sanctuary, who claimed the need to strengthen the security of its premises. The Data Protection Authority immediately ruled against this first request, and would only change its decision – albeit reluctantly – after the National Republican Guard drafted a well-substantiated project proposal. This proposal supported the need to monitor the site's entrance and exit routes in order to control the influx of thousands of pilgrims regularly converging there, and to help identify the high number of pickpockets who every year also seemed to make regular pilgrimages of their own (in the words of one of my interlocutors). But even then, recording was initially only allowed between 6:00 am and 2:00 am, though this was later extended to twenty-four hours a day due to the Pope's visit in May 2010.

The main argument involves the area's geographical features, by considering that a finer and more localized surveillance of the sanctuary's entrances and exits helps the police to have better control over the incoming and outgoing fluxes of population. The Data Protection Authority's report explains that besides the fact that 78 per cent of all the thefts registered in the village of Fátima were perpetrated in the sanctuary, with victims aged over fifty years, one of the main reasons why it recommends installation of surveillance cameras in this spot is as follows:

> The Sanctuary combines characteristics, some deriving from its own features and others from the activities carried out within it, which pose serious risks to security

> and public order, such as (i) the existence of only 6 exits, 2 of which are made through small tunnels with insufficient channeling capacity, and the remaining containing stairways . . . ; (ii) differences in terrain levels create the existence of large concave areas that make the inflow and outflow of persons harder; (iii) high influxes of people on specific occasions (concentrating up to 500,000 people simultaneously in the premises).[20]

This large agglomeration of persons (see Figure 2.5) which occurs in specific periods of the year requires a view of the location that allows the detection, for instance, of situations where there may be a need for first aid care intervention, as is often the case with pilgrims who frequently arrive in a state of extreme exhaustion after several days' travelling on foot.

Similar arguments were used in the project proposal for the Bairro Alto district, in Lisbon, but this time with completely different motives. Just like Oporto's Historic District, the Bairro Alto locale in Lisbon is one of the city's most active nightlife centres, attracting a great number of young people (especially on weekends) and tourists, but with it also some criminal activity. An urban layout of narrow, underlit streets that are filled with dense concentrations of people, added to a widespread consumption of alcoholic beverages, provides what is considered to be an ideal environment for criminal activity and disorderly conduct, such as drug trafficking, theft, burglary and street brawls.

Fig. 2.5: *The Sanctuary at Fátima: Detail of the Area for Burning Candles*

This circumstance, an old source of concern for local business owners, had for a long time motivated their demand for stronger security measures in the neighbourhood. In this instance, however, authorization was only granted for a trial period of six months and even then, monitoring was limited to night-time hours. Public video surveillance was defended largely by the local business association against the city mayor's and the local chief of police's initial opposition, to fight problems that had been clearly identified for years. In this particular case, video surveillance installation was only requested as a last resource, after other measures had previously been attempted. In fact, a different model had even been tried, but was shortly abandoned despite some degree of success at the time. Some months before Bairro Alto district's Commercial Association called for video surveillance, police presence had been strengthened in the neighbourhood, with 'extremely positive results', as he told me in conversation:

> At the time, the idea [of video surveillance] wasn't well received. Actually, the mayor's office even raised some privacy-related issues, which was understandable. District representatives, municipal police and Public Security Police members were present at that meeting, but only the head of the municipal police liked the idea. The matter was debated, with some differences of opinion, but overall nobody seemed enthusiastic about this solution; the main argument had to do with the protection of personal privacy.
>
> But we had previously gone through an interesting experience, when we – local merchants and the association – tried hiring paid policemen for a few months; they had special mobile phone numbers that could be reached by those who signed up. And it worked. Immediate action was needed to fight the drug trafficking carried out in plain sight on the neighbourhood's street corners. When the dealers saw the police, they would move to another corner, but then officers were warned by another business owner telling them 'they are here now', and dealers were forced to move again and again. Through this cooperation and interaction, dealers were at least forced to become less conspicuous. It is a whole lot of difference having half a dozen more or less hidden delinquents than the hundreds that crowded corners freely pushing drugs. That experience was very good. But it had to end because it was expensive; it cost us around 12 000 euros a month.
>
> We had become used to the opinion that there was no solution, and to fight this opinion, the movement was formed; and it succeeded in gathering a large number of supporters to pay for the police officers, which represented a significant monthly sum. But it was very efficient.

Unlike in the Baixa Pombalina, crime rates in this area increased in the period between 2007 and 2009, as the Public Security Police points out in the Data Protection Authority ruling, and added to the elevated cost that other methods seemed to imply, this eventually tipped the balance in favour of video surveillance for the Bairro Alto district. Despite being subjected to a six-month trial period (and not one year as had been the standard rule) after which it would have to be discussed again, this decision differs from that for the Baixa Pombalina, its neighbouring district, primarily due to the kind of activity that goes on there and the type of behaviour it fosters. Therefore, whereas in the latter case, where the high density of the population (particularly of the transient type) was not accompanied by an increase in crime, this fact actually worked against their project as an argument for requesting video surveillance – deserving the Data Protection Authority's negative vote – in Bairro Alto it weighed decisively on the final approval.

There were also other proposals, as the case of the project for Praia da Rocha in the Algarve, which caused greater public indignation and whose refusal seemed harder to explain. This seaside location is a popular tourist spot in summer. It is a small town, built around a central avenue, or strip, whose many bars and other nightlife venues become crowded by local and foreign visitors during the holiday season. Similarly to Oporto's, this project was intended to cover approximately 3 000 km^2, targeting precisely the above-mentioned area of bars. Even though Praia da Rocha has never registered high levels of crime, the local government and police forces believed that video surveillance would be a great aid in assuring local safety, hence they requested the installation of twenty cameras. Despite the similarities between these projects, the Data Protection Authority vetoed the one in Praia da Rocha on the grounds that the request did not make a convincing case. Data Protection Authority claimed that certain features in these devices might jeopardize the right to privacy and image in certain situations, given that their powerful zooming capability would allow them to reach far beyond the limited area of nightclubs, covering the whole beach used by bathers during the daytime.

What is evident so far is that with Law No. 1/2005, after years of utterly rejecting any request for surveillance cameras in public areas, the Data Protection Authority was for the first time put in a situation where it was not only forced to officially consider these kinds of proposals, but almost expected to approve them. As the proposals to install video surveillance for public areas progressed, the roles of the entities involved also became clearer and their positions more obvious. As a result, decisions concerning the authorization or denial of requests began to follow a pattern. While early rejections were interpreted as the natural result of reasonable caution, their persistence – little diminished by the already mentioned exceptions – exposed the Data Protection Authority's strong aversion to this particular device; but more importantly, its power of veto became more noticeable. As will be described in the next chapter, the Data Protection Authority started being portrayed as a blocking agent, harming the progress of strategically important reforms in the area of national and local security, and especially among the political elites, it came to be regarded as a nuisance and a problem that had to be solved somehow.

The pressure put on the Data Protection Authority had little effect, however. If anything, it seemed to entrench this organism even more in its convictions, making its decisions all the more categorical. On the other hand, new proposals seemed to ignore this evidence, and local entities continued to present projects that completely disregarded Data Protection Authority criteria. This cycle of misunderstanding continued to spiral, widening the already serious communication gap between the parties involved in this process. An example of this was the case of Batalha, a village with a population of approximately twenty-five thousand situated in the district of Leiria, a little north of Lisbon in the country's central region. This town is widely known for its medieval monastery (Mosteiro da Batalha), an imposing building of great historical significance, and classified as a UNESCO World Heritage Site. It is regarded as one of the country's national symbols and constitutes one of the region's touristic attractions.

In this village, the local government presented a project contemplating the installation of a single surveillance camera on the town's highest tower (coincidentally the National Republican Guard's headquarters), which would indiscriminately survey the whole town in a kind of archaic version of a panoptical Big Brother. This request was swiftly refused by the Data Protection Authority despite being strongly supported by the Ministry of Internal Affairs. Although this is an obviously extreme case, it is mentioned here to illustrate the levels of absurdity reached by the process at this stage.

The First Evaluation of CCTV in Open Areas

Thus far, the progress of several projects requesting public video surveillance, focusing primarily in the years of 2008 and 2009, has been described in a roughly chronological order. This narrative aimed to show how the emergence and interaction of different interests from various parties (chiefly economic and political) obeyed an often populist and circumstantial agenda. Even though this book's main concern does not involve a thorough scrutiny of the results that derived from the implementation of the cameras, this analysis will benefit from an account of the developments registered in these processes one year after they were enforced.

It should be remembered that according to law, after this period of time, the projects must be reappraised and go through the same procedures all over again: submission of reports from police forces, consent from the Ministry of Internal Affairs, and finally approval or rejection by the Data Protection Authority. The resistances and conflicts among the parties involved, in the period between 2005 and 2010 that is covered by the cases presented so far, will be the focus of the next chapter. Before that, there is a need to look further into the events that determined the fate of public video surveillance implementation in the cases that, due to the locations in question, bear greatest significance in the Portuguese context: Lisbon's Baixa Pombalina and Bairro Alto districts, Oporto's and Coimbra's historic districts, and Fátima's Sanctuary.

Let us start with the last one in the list, arguably the most successful case in the whole process. Nevertheless, even this case did not escape some early tension between the sanctuary's representatives and the Data Protection Authority, according to whom both the sanctuary and its surroundings were holy grounds for religious devotion, and as such, should be strictly protected against any type of image recording. However, after reviewing the report from the National Republican Guard, it was decided that religious freedom was not compromised since video surveillance did not reach inside the temple, being strictly used to monitor the site's exit and entrance routes, and thus becoming a great aid as an instrument of crowd control. To describe this case as successful is not to judge its effectiveness in terms of diminishing crime rates – which in any case the 15 per cent drop in criminal activity (namely petty theft by pickpockets) between May 2009 and March 2010 renders inconclusive – but merely a comment on its ability to provide this site with a security system in time for the visit of Pope Benedict XVI in May 2011.

The case of Coimbra is also relevant. This is a city of one hundred thousand inhabitants in the centre of Portugal – well known for its university, one of the oldest in Europe – where video surveillance was introduced in 2009, following the same reasoning as for the Baixa Pombalina; that is, in a relatively small commercial area, attracting a lot of tourism and consumers, twelve surveillance cameras were installed after some pressure from local shopkeepers who protested against what was described in the media as a 'crime wave' that had swept over the area, especially during the night-time. The cameras only operate in the hours between 8:00 pm and 8:00 am – the period considered more likely to see criminal activity – and are switched off in the daytime. The perimeter in this case covers a total of less than five kilometres.

In this city, as will be discussed further in the next chapter, the police did not consider video surveillance a particularly efficient method, especially since it implied taking officers from the streets and placing them behind cameras. Despite this reaction from the police and even the city council officials in charge of these matters, surveillance cameras were put in operation for one year, and the crime numbers presented by security forces for the renewal of video surveillance authorization, in 2010, showed a drop in criminal activity in several areas. These positive results, which the Ministry of Internal Affairs did not cease to emphasize in the media as a public demonstration of the measure's success, nevertheless seemed to omit or ignore the real situation. In fact, by the end of 2010 criminal figures had risen again, and the area was plunged into the same problems it suffered before the cameras had been installed.

One of the arguments that most persuaded business owners and the public in the early stages of the process had to do with house and business break-ins in the area at night. At the time, the Public Security Police considered that burglaries and acts of vandalism (broken shop windows for instance) registered in this area could not be regarded as being more serious or threatening than similar situations in other areas. According to the Public Security Police officer responsible for the project, even though the city's historic district did not register a greater amount of crime than other commercial areas, it was thought that the video surveillance devices might have the placebo effect of reassuring its population.[21] He would see in later events a confirmation of the scepticism conveyed to me at the end of 2009: cameras did not prevent or diminish crime, they just made it move temporarily, and neither did they deter criminals nor make police action any swifter. In November 2010, 'Serial Crimes in Coimbra' would make the news again, in an article that read:

> Despite official Public Security Police statistics showing a decrease of crime in the area compared to last year, the fact is that the issue will once more be discussed at the mayor's office between representatives from the Coimbra downtown Business Association and the police.
>
> The same problems are once again invoked by the president of the merchants' association regarding the area's urban planning: 'It is a difficult area to manage, with many narrow streets, mostly deserted at night; we know that we can't have a police officer at every corner. We have already also realized that video surveillance (monitored from the PSP downtown headquarters) by itself is not enough'. (*Diário de Notícias*, 10 November 2010)

During the last trimester of 2009, two other requests for public video surveillance were ruled on by the Data Protection Authority, both for legendary areas of Lisbon: the Baixa Pombalina and the Bairro Alto district. The former was rejected and the latter approved. A comparison between these two cases is relevant since it exposes some contradictions inherent to projects of this kind in Portugal, at least in the period under consideration here. As described above, the project for Baixa Pombalina was submitted on the parish president's initiative and apparently had everything in its favour. However, the idea for installing thirty-two cameras to monitor a district commonly considered the heart of the city was bluntly refused by the Data Protection Authority. There was an immediate reaction and a new project was submitted in 2010, this time with the mayor's intervention, strong support from the Ministry of Internal Affairs and the cooperation of the Public Security Police. In the following meetings I had with members of these organizations, it was interesting to see the different positions unofficially assumed by each party.

The Ministry of Internal Affairs Secretary of State mostly expressed her objections to the role played by the Data Protection Authority, which in her opinion should stick to its duties and withhold from ruling on the validity of using cameras. In other words, her opinion was that the Data Protection Authority should be a consulting entity and its rulings without binding force. The mayor of Lisbon, in his turn, led a public campaign to defend the relevance of an even more extended version of this project that would include fifty cameras instead of the initial thirty-two. His position, apparently contradicting the Data Protection Authority's previous ruling – remember that the high number of cameras had been one of the reasons pointed out for its refusal – was explained by an enlargement of the geographical area to be covered, as well as by the types of crime presented by the Public Security Police for these sectors. In this second project, the area covered was not limited to the streets of the Baixa, with their commercial avenues, touristic and leisure sites, but extended to the adjacent, less distinguished quarters. This included Martim Moniz Square, where there is a high concentration of retail businesses run mostly by Bangladeshi immigrants, and Intendente Square, an area of Almirante Reis Avenue, traditionally known for prostitution and drug trafficking.

One of the main mottoes was 'rehabilitate', and as the first sign of this change, this will to make these areas safer and cleaner, the mayor's offices themselves were to move to this square. Public Security Police added new and different types of crime to their previous report, namely the so-called cases of 'incivility' which caused social alarm, such as window smashing or graffiti. This new project, drafted in December 2010 but only publicly announced in January 2011, was once again rejected by the Data Protection Authority, whose ruling states: 'Incidentally we observe this new request's apparent disregard for the previous negative ruling, making no effort to solve the legal problems raised thereupon'.[22]

In the case of the Bairro Alto district, despite the repeated announcement in the media of authorization for surveillance camera installation, by the end of 2010 not a single concrete step had been taken in terms of launching a public competition for the purchase of cameras and other necessary equipment. That is, more than one year after the assessment of this area had been completed – which, as previously discussed, included the intense night-time activity and high concentration of bars, dense crowds of young people especially on weekends, presence of criminal activity that required strengthened security measures, etc. – the situation remained unaltered, and without any prospect of change in sight. As a news piece from *Diário de Notícias*, 15 December 2010 reported, Lisbon's Civil Protection alderman[23] stated that 'besides the lack of experience in budgeting video surveillance systems, since it is the first of its kind in the city, the municipality faced technical problems with the installation of the fibre-optic cable, made worse by the fact that it is a historical area containing listed buildings'.

The real obstacle in this case, however, cannot be found in the inexperience of Lisbon's city council employees to draft a budget, given that although video surveillance was new to this city, in this particular detail they could have drawn on the experience of the Oporto project or even the one drafted for Coimbra, since both had similar characteristics. As in other aspects surrounding the installation of public video surveillance, the lack of consensus and dialogue between intervening parties did on the other hand become obvious in a process that was supposed to be a joint operation.

Oporto is perhaps the best example of this general disjointedness. It was a troublesome process from the very beginning; from the initial divergence of opinions between the president of the Commercial Association and Oporto's mayor, and later with the granting of a partial authorization that did not meet the area's needs, to the various confusions as to who was responsible for financing the building of infrastructures and paying the cameras' energy bills. The result was that the cameras had been in place, but not working for a whole year before Oporto's Historic District became effectively operative at the end of 2009. The Public Security Police's early claims that cameras were of little use, given the area's scarce criminal incidents, did not prevent the Ministry of Internal Affairs from reporting the 'excellent' results of this security device in 2010 widely in the media, supported by statistical data provided by that same police force.

In spite of this apparent success, the report submitted from the Data Protection Authority at the time of renewing authorization for this project revealed that: 'not only does 'the presence of the system [does] not seem to have had an effect on crime rates, as in fact those types of crime (robbery, theft) where a clear decrease would be expected have experienced an increase since video surveillance started'.[24] On this point precisely, it is worth attending to the words of a police officer speaking to me about one of the places where public video surveillance had been operating since 2009:

> We have now reached the end of the first year and all we can conclude is that police activity in this area has been close to none, but in itself this does not tell us anything about the merits or demerits of video surveillance; there are two possible interpretations: either we conclude that the low crime rates in the area indicate that video surveillance is unnecessary, or we can speculate that these numbers are a result of its presence.
>
> You simply cannot measure a non-occurrence; that is precisely what pro-activity means. Whether the presence of surveillance cameras in itself prevents criminal activity, we will never know. At the most, we can make an interpretation based on the comparison with previous crime rates, but even this is a purely conjectural reading that is ultimately based on false premises, if for no other reason than the fact that the motives driving those persons are not necessarily the same as the police's.

The project for renewal, submitted in November 2010, requested an extension of surveillance to a 24/7 monitoring regime, which would again be denied by the Data Protection Authority, based on police reports showing that criminal activity was concentrated in the night period that was already being covered (between 9:00 pm and 7:00 am). Besides this decision from the Data Protection Authority, the end of that year registered a new setback, this time coming from an unsuspected source, the system's main campaigner: the president of the Commercial Association himself. António Fonseca stated that the cameras should be simply turned off because they were not worth the three thousand euros that the association had to spend every month on them if they could not operate for the twenty-four hours a day considered necessary by this entity for them to be effective.

What becomes interesting when the evolution of the process in this particular case is observed, is that from a certain point onwards the whole project, which had started on Fonseca's initiative, ended up also becoming dependent on his actions regardless of other parties' intervention, except of course of the Data Protection Authority. It was the president of the Commercial Association who raised the problem of poor security – despite not being backed up by the police – and forced this process to move forward. It was also up to this same president to fight publicly for this initiative, including contesting the Data Protection Authority's partially positive ruling. Lastly, he was also the one to decide, after the media strongly protested against the Data Protection Authority's refusal to grant an extension of surveillance activity in the area, to turn off the cameras – which effectively occurred on February 2011 – and have them removed from their location. Neither the Ministry of Internal Affairs nor local police forces assumed an active voice in these procedures or moved to reverse the situation in order to preserve the cameras, even though they were legally entitled to, seeing that the Data Protection Authority had not actually denied the continuation of video surveillance in the area.

Summary

This chapter has given a detailed description of public video surveillance projects between 2005 and 2010, paying special attention to the progression of events and their outcomes. Besides the differences in the final results, it has tried to highlight the distinguishing features of each case, while emphasizing an aspect that has to do mostly with the high expectations created by this project as a whole. Therefore, the next chapter will no longer be concerned with events themselves, but with the more abstract question of how the interested parties rationalized and interpreted this process. As will become abundantly evident further on, the underlying motivations in this whole process (whose political character contains a strong economic element) are substantially more intricate than a superficial reading might suggest.

Notes

1 On this point we might recall Pina Cabral's comments on the 'dynamics of identity' applied to the context of video surveillance in postmodern society: 'Anyone who initiates an action of surveillance intervention must always assume that, for each of its immediately foreseeable effects many other will arise which have not been anticipated. . . . One of the biggest risks in contemporary surveillance is to let oneself be tricked into thinking that it is possible to have absolute control over the uses given to the information one has gathered (to the objectifications one has produced). In the absence of evidence that proves otherwise, one can never presume that there will be no corruption, inefficacy, human weakness, greed or impudence' (Pina Cabral 2008: 25).

2 Amongst several works on Foucault, see for example the volume edited by Jonathan Xavier Inda (2005).
3 See the special issue from 2003 of the journal *Surveillance and Society*, dedicated to 'Foucault and Panopticism Revisited', which includes several articles exploring the connections between Foucault's theories on the panopticon and current surveillance practices.
4 For a thorough discussion of Agamben's 'state of exception' in connection with the works of Hannah Arendt and Michel Foucault, see Douglas (2009).
5 There is a vast literature on the use of video surveillance systems in 'exceptional' circumstances such as mega-events. See, for instance, Klauser (2009a); Bennet and Haggerty (2011).
6 *Diário da República*, Série A, no. 6, 10 January 2005, pp. 205–208.
7 Ibid.
8 Ibid.
9 Ibid.
10 Http://www.mai.gov.pt/data/legislacao/Videovigilancia2.pdf
11 We must nevertheless allow for the time it takes for this kind of procedure to go through – roughly one year in this first case – and also bear in mind that an official go-ahead does not necessarily mean that an area is immediately put under effective video surveillance.
12 The population of Oporto comprises approximately 220 000 inhabitants; the capital city, Lisbon, doubles this figure with a population of 560 000.
13 The quotations cited here are included in the Data Protection Authority's final decree on this case, namely Decree 60/2007.
14 Ibid.
15 Ibid.
16 In a sense, this is in accordance with Goold's analysis of the exponential increase in the use of CCTV in public areas. He says that we should 'regard the growing use of public surveillance by the police as part of a more general trend towards what has been called 'risk-based policing'. According to this view, collecting and disseminating information that can be used in the assessment and management of risk has become one of the core functions of the modern police force, progressively replacing more traditional concerns about law enforcement and crime control (Goold 2004: 4).
17 This rationale closely resembles Caldeira's observation of the existence of 'a widespread aesthetic of security' (Caldeira 2000: 257) that confers prestige to new types of condominiums in São Paulo, Brazil. According to the author, reinforced security – fences, armed guards, internal CCTV etc. – actually becomes an element that confers status to its inhabitants.
18 See Data Protection Authority Ruling 62/2009, where these quotations are displayed, www.cnpd.pt.
19 Ibid.
20 Data Protection Authority Ruling 36/2008, p.6, www.cnpd.pt.
21 A newspaper article (*Jornal de Notícias*) dating from 16 January 2010 on video surveillance in Coimbra reads: 'Merchants and local leaders agree that the video surveillance system installed downtown has brought greater calm . . . . Luis Quintans, an antique shop owner, claims that the cameras that are in place "give people a psychological sense of safety. Personally I don't feel any difference, but I think the atmosphere is calmer"'.

22 Ruling 5/2011, from 17 January 2011. www.cnpd.pt.
23 Civil Protection (*Protecção Civil*) is the equivalent to the English Civil Contingencies Secretariat, 'with the aim of preventing collective risks inherent to serious accidents or disasters, to attenuate its effects and to protect and help people and assets in danger, whenever those situations occur' (http://www.prociv.pt).
24 Data Protection Authority, Ruling 2/2011, 10 January 2011.

Chapter 3

Policy-Making

Successes, Failures, Contradictions

The previous chapter situated the theme of video surveillance and its enforcement within the European context, while providing a detailed chronological account of the implementation of this device in Portugal in the period between 2005 and 2010. It is clear that this was a troubled process, as much in terms of its aspirations as in terms of the roles played by the different entities responsible for conducting and executing it. Measures deemed imperative by the Ministry of Internal Affairs were sometimes trivialized by police forces, apparently unwilling to admit that video surveillance was actually useful, or in any case reluctant to follow advice from external organisms on the proper performance of their own assignments. In addition, despite the almost unconditional support of the ministry's representatives – who incessantly encouraged projects and approved of them – the fact remained that these did not always conform to the law, insofar as they did not meet some of its requirements regarding legitimately established goals and purposes.

The progress and outcome of consecutive requests revealed yet another obvious fact: the position of the Data Protection Authority, whose power of decision was legally binding and final, emerged as being in principle against the use of such devices, especially in open areas. As the research developed, each of these actors' positions vis-à-vis the use of public video surveillance (even if not always assumed as such officially) gradually became more evident. An analysis of the details for this period concerning video surveillance proposals and their implementation in Portugal also revealed the diversity of proponents and specific objectives: some originated in professional associations interested in protecting their businesses or areas of business; others were initiated by councils which spontaneously advanced proposals of their own, later seeking the cooperation of local police forces, on whose official reports the proposals' designs were ultimately based.

This chapter will strive to demonstrate that the political initiative to introduce surveillance cameras in open areas in Portugal mimics the standardization of similar practices within European countries, although apparently without the same degree of success. It was motivated by the need to correspond to what may be called a European video surveillance model, and translated as a political response which sought to show that action was being taken to solve a very specific problem, namely the alleged feeling of insecurity expressed by populations. However, this governmental strategy for safety and security was being enforced in a country which, according to police estimates, had low levels of violent criminality, no risk of internal threats such as regional independence movements or any sort of ethnic problems or social instability, and finally, a geo-political alignment (especially as part of NATO and involved with their foreign policies) that meant that risk of external attack was never seriously considered.

It is important once again to stress that the great boom in public and semi-public video surveillance in the United Kingdom, which is also the major reference point of the parties involved in the Portuguese project, occurred at a time of internal economic crisis, with reflections on the rise of violent criminality and accelerated degradation of vast urban areas. The political and economic strategy to regenerate those spaces was directly linked to this wave of criminal behaviour. Even so, the dissemination of these systems in the U.K. was so fast and widespread that looking back we can appreciate how there was almost no prior critical reflection or careful assessment of which areas were subject to video surveillance. Instead of this thorough prior diagnosis, there was an almost uncontrolled diffusion of CCTV, installed without any planned or apparently understandable criteria. In other words, the reasoning would be something like: 'if CCTV is useful to prevent and fight crime, it should be used without restraint since every place could be the object of unlawful (and dangerous) behaviour'.

Almost two decades later, and after several studies demonstrating the negligible effect of video surveillance in terms of crime deterrence and prevention (notwithstanding its unquestionable value as an instrumental aid for criminal investigation), some municipalities have started to ask themselves whether the investment involved maintaining thousands of cameras still makes sense, or whether it is time to review this policy and revert to alternative means (Doyle, Lippert and Lyon 2011; Webster 2009, 2004). In any case, the fear of crime that the massive and frenzied reaction witnessed in the British case differs substantially from the perceived feeling of insecurity under discussion in the Portuguese context.

Considering some discrepancies about the motives, main goals and even official positions regarding the use of CCTV in public areas in Portugal, and keeping in mind that the main reason to use this device is to act directly to combat the feeling of insecurity – by itself an ambiguous notion – it may be asked: is this a case of pure political dispute between parties, government and institutions using the leitmotiv of security as a pretext to maintain and somehow affirm their own power, in a struggle that is all but successful in terms of its initial and allegedly primary goal? To answer this question and what it implies, in this chapter I will pursue a line of reasoning that is very close to Jonathan's Spencer when he proposes that:

> Rather than imagining the relationship between 'the state', its projects, and the world, as one of the gradual actualization of a blueprint, the making of the world in the image of a certain rationality, we could start from the central ambiguity of statecraft, its failure to produce a world, or subjects, or spaces for social or political action: rather statecraft is better seen as performative, it attempts to enact a version of the world against a background of dissonance and transgression. The border is where the state performs its rites of order and control. (Spencer 2007: 116–17)

The art of leadership and command over a country – in other words, the art of governance – is described by Spencer as 'statecraft', a skill that is constantly performed in the space between action and intention, between pronouncement and fact, between reality and its representations. In that sense, institutional performance is as much a producer as a product of those virtual spheres where the battles – whether political, institutional or both – are fought.

I have already alluded here to the notion of meta-discourse and ascertained the relevance of the concept of modernization as a pretext for policy-making. The case of public video surveillance presents a similar formula. The discourse that bridges intentions and actions becomes all the more effective as it is repeated in the public arena, a connection that in this case is established between politicians and the population through the media, additionally becoming more entrenched in its convictions as it is met with opposition. It would therefore be wrong to restrict the idea of installing video surveillance in major urban areas to a matter of crime-fighting policies, ignoring policy-making in the broader sense, and what I consider to be the problem of its meta-discourse. To explain this notion, I turn once again to Spencer, and his observation that it is crucial 'to pay attention to what people say they do (or what official documents say they should be doing), and also to what they actually do. And, noting the inevitable discrepancy between the two' (Spencer 2007: 116). According to the author, it is the discrepancy itself which becomes meaningful. Thus, understanding the whole phenomenon 'involves a combination of wide-eyed empiricism – looking at what is actually there – and the most critical and suspicious of interpretations' (Spencer 2007: 116–17).

Roughly speaking, this is how it operates: a given initiative is conceived with a particular purpose, and as soon as it is announced it begins to have effects, even when it is not carried through. This does not amount to deception or dissimulation, but is simply a case of meta-discourse, of pronouncing intentions and goals that do not always have to be materialized, as Pierre Bourdieu so aptly described thus: 'in politics, "to say is to do", that is, it is to get people to believe that you can do what you say' (Bourdieu 1991: 190).

Notwithstanding the diversity of motivations driving the different entities involved in the Portuguese case, it is crucial not to lose sight of the legal framework that enabled it – first and foremost the general law which regulates public video surveillance – and the governmental plan responsible for conferring to it a national scope through the National Video Surveillance Programme. It has already been stated that video surveillance in public spaces in Portugal emerged in the context of a political design bent on introducing advanced technology in all areas of state activity. It must therefore be understood as an instrument of modernization (one amongst many) intended to hoist the country to levels of development that seek to emulate the models of more advanced European countries. This aspect of permanent comparison and imitation becomes particularly notorious in the area of video surveillance in open areas, following the kind of simple reasoning that can be summed up as follows: if cities in France, the United Kingdom, Italy or The Netherlands use these monitoring devices to aid local crime fighting, why should not the same technology be used in Portuguese cities? The undertone to this proposition being: how are Portuguese cities any less important than their European counterparts are, in terms, for instance, of national and international tourism?

However, as soon as the more objective questions of video surveillance's effectiveness in strengthening personal and material security at a national scale enter the discussion, the different interested parties start contradicting each other, and the lack of strategic coordination and harmonization of goals – even in terms of their interpretation of Law No. 1/2005 – come to the surface. At this level, alas, no one seemed to agree on the purpose of or expected the same results from video surveillance devices. This transition from the ostentatiously unanimous ambition of a national design, embodied in the National Video Surveillance Programme, leads one to wonder what might have possibly gone wrong. One does not have to dig deep to find the first blatant clue to this. Apparently, the official proposal for this project, which intended to cover the entire country with video surveillance devices, formally consists only of eight short points, mostly contemplating major highways and roads. Just one of these points actually specifies the usage of surveillance cameras in urban areas, ultimately the crucial point at stake here. In fact, not even this programme's nationwide scope was consensual, as a careful analysis of the discourse maintained by the different intervening parties in the various geographical areas reveals.

The following sections will strive to shed light on the initial stage of video surveillance implementation in Portugal. This will be crucial to understanding any future developments, inasmuch as they may shed light on the original motivations and dynamics involved (accounting for the evolution of power struggles and how they began to be resolved; see Herzfeld1992) and map the learning process behind the consecutive project proposals and their resultant rulings. Each of the following sections is dedicated to one of the major interlocutors engaged in the elaboration and decision process of installing public video surveillance systems: the Data Protection Authority, police forces, government authorities, and other political bodies. The last part of this chapter combines all these elements while establishing some sort of dialogue with works that analyse similar political processes in other parts of the globe, this time exclusively focused on modernization and development.

The Data Protection Authority

The Data Protection Authority 'is an independent body, with powers of authority throughout national territory. It is endowed with the power to supervise and monitor compliance with the laws and regulations in the area of personal data protection, with strict respect for human rights and the fundamental freedoms and guarantees enshrined in the Constitution and the law'.[1] It is composed of 'the Chairman and two members elected by the Parliament; one legal magistrate, with over 10 years' experience, appointed by the Magistrate Council; one magistrate from the Public Prosecution Office, with over 10 years' experience, appointed by the Public Prosecution Council; Two members appointed by the Government'.[2]

Besides its other responsibilities in matters of data protection, this entity has a power of decision over the installation of surveillance cameras in open areas that is legally binding. This capability was granted by law, and in terms of public video surveillance, more specifically by Law No. 1/2005, which decrees that: '1) – The present law stipulates that the installation of fixed cameras must be authorized by the member of government in charge of the proponent security force or service after being ruled by the Data Protection Authority (DPA); (2) – Authorization cannot be granted in case of negative ruling by the DPA.'[3]

This binding force demonstrates that the legislator, at the time that the law was conceived, wished to entrust solely to the Data Protection Authority the ability to determine – based on an appraisal of the arguments and information presented to support a given video surveillance project – whether the rights of those affected by it (as far as personal data, privacy and intimacy are concerned) were in any way being violated. This is a good time to stop and consider some of the preoccupations expressed by this entity on such matters. The Declaration proffered on 28 January 2010, on the occasion of the European Data Protection Day, bluntly lists the risks that in their opinion citizens are currently being exposed to – often willingly – also raising some objections to potential misuses of personal information, considering a world where information and communication technologies are routinely used by people of all generations for infinite purposes.

> The celebration of European Data Protection Day gains special significance this year with the entry into force of the Lisbon Treaty, which enshrined the protection of personal data as a fundamental right of European Union members. This is undoubtedly an important event in the deepening of European democracy, particularly at a time when it is increasingly being imposed on individuals to relinquish their privacy and their freedom in the name of public or corporate interests.

> We must highlight that Portugal was a pioneer as the first country to recognize data protection as a fundamental right in the Constitution of 1976. With the establishment of democratic rule, it sought to reassure citizens that their rights and freedoms were safeguarded from information technology, whose potential for application was evident even then.
>
> In fact, over the past 30 years, technological developments have not ceased to amaze us, reaching levels that are truly admirable, both by the rapidity of their progress, and by the extent of their achievements, bringing undeniable benefits to people's lives and societies. The information and communication technologies especially, have radically changed the world as we knew it, giving the phenomenon of globalization an unprecedented substance.
>
> However, this same technological capability has also permitted the creation of large, often interoperable, information systems, crossing millions of personal data at an increasing pace. Technological and economic synergies raise important questions regarding intrusion in the privacy of each and every one of us. It is with grave concern that we have observed an accentuated trend to collect increasingly more personal information on citizens, to control their movements, to know their habits and preferences, to monitor their individual choices.
>
> The profusion of biometric, video surveillance and geo-tracking systems, the large-scale use of registered Internet activity, the elaboration of detailed individual profiles and the resulting discriminatory labelling of people as well as the proliferation of black lists and indexes are symptoms of a monitored society, on the path to substantial control over the individual.
>
> Moreover, the mass treatment of personal information is often carried out in such a non-transparent manner that it becomes almost invisible to people. This scenario has developed from a disturbing compound of security reasons, in the name of which the use and convergence of surveillance technologies have been enhanced, causing the majority of citizens to be judged based on the premise of suspicion.[4]

In this Declaration, the Data Protection Authority emphasizes two major concerns. The first one has to do with the potential uses of information and communication technologies conceived to gather personal information: namely their profiling effect in terms of individual 'habits and preferences' – information often provided by consumers themselves. This is used, for instance, in corporate consumer profiling aimed at establishing target markets, or to distinguish between attractive and unattractive clients.[5] From this declaration's reference to personal data and their potential misuse, one other significant aspect can be inferred. It is connected with the idea that citizens are generally unaware of the consequences, responsibilities and overall implications at stake, which is tantamount to claiming that the notions of civil rights and freedoms are virtually absent from common perceptions of this subject.

It has already become obvious that this declaration is much more than a merely informative document, and has a clearly political intention. It conveys in no uncertain terms this entity's conviction that recent times have promoted a generalized atmosphere of alarm – specifically regarding people's fear of their own physical safety – which explains the implementation of technological systems as an alleged counteraction. This statement points to a conclusion that in my opinion underlies many of the arguments presented in the next pages; that the dividing line between security and privacy, fear and freedom, is indeed a very thin one, if at all existent. The Data Protection Authority openly criticizes policies that seek immediate responses to fear – almost a reflex reaction to this phenomenon – instead of previously making a sustained assessment of whether they are actually the most suitable means to guarantee a balance between security and privacy. In fact, the declaration's concluding remarks define the Data Protection Authority's main goal well:

> It is urgent to reflect upon the path that is being followed. We cannot continue to feed the fear of people and make it easier for them to waive fundamental rights. It is always necessary to heed the principle of proportionality in order to find the best solutions.
>
> It is essential that before making choices and executing decisions which might have reflections on people's rights, a systematic and integrated set of studies is carried out, through which their impact on privacy may be assessed and the consequences on the lives of citizens be considered.
>
> The Data Protection Authority is fully aware of present and future challenges facing data protection and privacy, both nationally and at European and international levels. The protection of personal data is not just a personalvalue, but also a collective good that must be defended.
>
> It is not an easy task and requires the awareness and committed participation of all: the state, business, media, schools and every one of us, collectively and individually. It is a duty which on our part will not be evaded.
>
> The failure to safeguard privacy calls into question other rights and freedoms such as freedom of expression, freedom from discrimination, the right to free movement, the right to anonymity, and the limits of human dignity.
>
> A democratic society implies a constant fulfilment of the necessary conditions for the effective exercise of fundamental rights and freedoms. The Data Protection Authority hereby reaffirms its commitment to defend data protection and privacy.[6]

Overall, this document touches upon a variety of issues, such as the use and protection of personal data, consumer habits, control over individuals, the use of biometric data (such as fingerprints or iris pattern recognition) or the regulation of virtual space, in an attempt to encourage discussion and reflection upon the consequences of revealing personal information in today's world.

The principles hereby invoked, which are in accordance with those set down in the Constitution of the Portuguese Republic, convey the principle defended by the Data Protection Authority regarding the protection of personal information. But neither its principles nor its decisions are always compatible with the positions of other parties. Subsequent negative rulings for projects of public video surveillance have revealed this entity's conservative position in terms of its tacit acceptance or support of the use of these security technologies (which on the contrary it views as something that could be called 'insecurity technologies') by political power or state police organisms, thus somehow assuming towards these parties the role of regulator in terms of the protection of civil rights and freedoms.

Ironically, persisting with this position over the years has not brought the Data Protection Authority much popularity, and this is not just due to what was commonly perceived as their boycotting of public video surveillance. There were several instances in which the Data Protection Authority was publicly criticized for following its own agenda (considered by many as excessively conservative if not outright retrograde) regarding, for example, the use of genetic data for purposes of medical and scientific research, even when financed and conducted by organizations theoretically considered suitable and their purposes legitimate.

This point is significant because it allows us to appreciate the power retained by the Data Protection Authority, which was legally and constitutionally bestowed on it. Regarding video surveillance, in view of the fact that the law was built so as only to concede temporary authorizations for the working of these kind of systems, which must additionally be reviewed annually, this agency considers that such these devices should never be used systematically or become a permanent response to crime, as the National Video Surveillance Programme apparently implies. In short, the Data Protection Authority questions the appropriateness of surveillance cameras as a crime-fighting instrument in the hands of police forces when weighed against the rights of thousands of citizens potentially affected by it, particularly regarding their rights to image and free circulation – their right to carry on their daily lives free of constant monitoring.

In many of its rulings, the Data Protection Authority suggests alternative solutions, either in terms of different aids to policing, both through the strengthening of human resources and through additional material means, or other urban planning measures such as better public lighting and rehabilitation of the more seriously decayed areas. Even so, political forces often regard these suggestions as extemporaneous and beyond the competence of the Data Protection Authority, from which they only expect a simple ruling based solely on the data provided, disregarding any other information.

This leads to another interesting question, that of the surprise shown by governors towards the Data Protection Authority's interpretation of the law and its rulings on video surveillance. In fact, the Data Protection Authority believes that the legal framework in which Law No. 1/2005 operates still contains legal gaps, and that its interpretation contradicts the principles that supposedly guide the National Video Surveillance Programme. Thus, it becomes clear that the enthusiasm with which the Ministry of Internal Affairs welcomed the requests for video surveillance between 2005 and 2010 was not similarly expressed by the Data Protection Authority, and in addition, that the programme did not result in a coordinated strategy of the different state organisms.

If we consider the original intention of Law No. 1/2005, and how it was interpreted by the Data Protection Authority – 'requiring proof of the existence of objective risks' to continue surveillance over those areas for which the system is intended – we can see that it essentially demands the presentation of concrete data; in this case, those provided by police crime statistics. However, where criminal conduct and illicit behaviour are concerned, the essentially objective or quantitative character of this data may sometimes prove inadequate to provide an accurate portrayal of a given situation. Criminal occurrences are frequently not officially reported, either because victims decide that it is just too much trouble to go through the time-consuming bureaucracy over stolen items that are ultimately perceived as relatively worthless, or because that sort of common crime is usually hard to prove. As a detective from the criminal investigation unit of the Lisbon Public Security Police told me: 'It seems incredible, but people are so scared that frequently they are unable to provide even a basic description of the mugger'.

The discrepancies between the objectivity of crime rates on one hand, and the inherent subjectivity of personal feelings – whether those expressed by victims or those expressed by people who simply hear about these occurrences – on the other, make it hard to accurately assess the true dimension of criminal phenomena. In my opinion, the Data Protection Authority to some extent exploited this discrepancy, using it as a pretext to rule against projects for the installation of public video surveillance.

The situation so far could be described as a kind of catch-22. The Data Protection Authority acts simultaneously as the guardian of collective and individual interests, and as a *sousveillant*, making use of all the restraining measures in its power to monitor those who seek to do active surveillance. Since by definition the law must be equally and uniformly applied, ignoring differences between cases and contexts, it allows no exceptions based on unquantifiable elements, as for instance one might classify the feeling of insecurity. Thus, a final analysis based solely on numbers – which, it should not be ignored, are paradoxically easy to manipulate and/or interpret subjectively – can at best be understood as a half-truth.

The population's feelings or concerns, as I was told at the Data Protection Authority, are considered ambiguous or beyond the spirit and the letter of the law, and as such become irrelevant. The law must be equal for all and applied indistinctly. Any attempt to interpret society will be 'riddled with inaccuracies, omissions, faulty aggregations, fraud, negligence, political distortion, and so on' (Scott 1998: 80). From another perspective, while it seems clear that the presentation of criminal statistics by the police is mandatory to ascertain levels of danger and insecurity in a certain area,[7] we can also see how the Data Protection Authority uses this same material to question the very authority that, in turn, is trying to implement video surveillance in open areas.

The cases presented in the previous chapter show that the process usually goes through the following chain of events: (1) the population complains about criminality and insecurity; (2) politicians and the media explore these expressions of concern, creating an image of what seem to be unprecedented levels of danger; (3) when the security forces present their reports and statistics, the Data Protection Authority considers whether the numbers have real or valid significance. As members of this authority will admit, they make a very strict interpretation of the law that sustains a truly conservative opinion with regard to monitoring large public places, especially when it implies national coverage.

Police Forces

> Surveillance cameras are here and they are operating. If you're asking me whether I'm glad to have them, I would definitely say yes; any device which may help the police do their job is welcome – I would be equally satisfied to have more police officers or patrol cars.
>
> But if you're asking me: 'Do you need the cameras?' then I honestly couldn't be so sure. I have never put it like that, but I would have to say no – ultimately I am following orders. I have never publicly defended what I am telling you here, but the truth is that I have never been asked before. (Superintendent in charge of the surveillance cameras project in the Oporto Historic District)

Another agent playing a decisive role in the process of requesting public video surveillance is the police, the authority that holds more weight when it comes to either assessing the need for this mechanism and its potential uses or establishing the geographical boundaries in which it will operate. In order to understand fully how police and other security forces interpret video surveillance projects and even to comprehend their overall reaction to this phenomenon, one should first remember the prominent role played by the political police (PIDE) during the Salazarist regime. Albeit created with very specific purposes, namely to capture and control political opposition, this organism counted on the support of the whole security apparatus, including other police bodies such as the still existing National Republican Guard and the Public Security Police. Notwithstanding the important part played by the latter in the period immediately after the 25 April coup and in the following decades, and even though the more repressive branches of the old regime – such as the PIDE and the Portuguese Legion – had been totally dismantled after the revolution, these institutions continued at some level to be burdened by their historical legacy, as a symbol of totalitarian power.

The current discourse of these institutions' members regarding their position on public video surveillance bears witness to the relevance of this historical legacy and its significance for this study as one of the instances where the persistence of popular suspicion towards authority has most strongly endured. In his study of the restructuring process of police forces in their transition from dictatorial regimes, and his comparative analysis of the Portuguese and Spanish cases, Palacios Cerezales has shown that in Portugal, despite the changes occurred immediately after the fall of the old regime, 'the civil rights frame did not recognise the existing police as a member of the new democratic polity' (Palacios Cerezales 2010: 434). In fact, as Cerezales also explains, due to their active participation in the dictatorship's repressive policies, the power of police forces as representatives of the 'law and order frame' was drastically affected, especially during periods when any initiative on their part to enforce public law and order was immediately connoted with fascist tendencies. Cerezales even describes the recognition that the police's authority was fragile as a traumatic experience, a phenomenon whose effects, in my opinion, persisted most acutely within police forces themselves (particularly the Public Security Police and the National Republican Guard) until at least the end of the 1990s. As mentioned by one officer from the Public Security Police:

> [Video surveillance in public areas] could never have been the Public Security Police's idea, and was always internally met with some reserve, because we believe that the police must not take the lead in such a controversial matter.
>
> In everything that has to do with pioneering police techniques, the general idea is that police forces must always follow behind the social attitudes of the moment. Particular innovations are always met with scepticism whenever they are initiated by police forces. It is a little like the common idea that the police is always one step behind; this is something we really have to get our heads around, and sometimes rightly so – that we must actually adjust to events.
>
> The reason is simple: in any situation which may involve an alleged violation of privacy, excessive use of force or disproportionate measures we are immediately branded as authoritarian.

The structural transformations, in terms of both the image society had of police forces and these forces' actions at the service of the state and its citizens, occurred as a slow and gradual process. New police training programmes intended to raise levels of professionalism and competence within these institutions were only introduced at the end of the 1990s, with the opening of higher education police academies and their focus on specialized training, once and for all abandoning the military element in their structure. As Palacios Cerezales explains, in the case of Portuguese police forces there is a 'double legacy' of authoritarianism. On the one hand, it is linked to the Salazarist regime, insofar as they were unable to deal with the changes produced by the democratic transition. On the other hand, it has to do with resistances found in their own internal structure as well as with the image projected by police forces to those outside. A deep remodelling of the whole organization would only effectively start to occur near the end of the 1990s, slowly transforming that image of backwardness and resistance into one of democratic values.

Given that video surveillance in public areas is the exclusive responsibility of state security forces, as opposed to in semi-public locales, where it may be carried out by private security companies, the National Video Surveillance

Programme was publicly announced by the government to aid the police in the war against crime and reinforce their current security devices. But despite the police officers' apparent acceptance of the surveillance cameras and their cooperation with the project, this process did not originate from their own initiative and would never have done so, according to the officers with whom I spoke. On this point they agree with Susana Durão's observation that police agents 'find themselves caught between political, organizational, communitarian and social dynamics. . . . All activity is conditioned by a balance of power between often contradictory internal and external forces which apply their pressure at different levels and with varying degrees of exigency. The state and the seat of power are a less unified force than they are generally held to be' (Durão 2008: 17–18).

As a state security force, the police have to harmonize, within its own guidelines, political initiatives *stricto sensu* – that is, conceived away from the field – with the specific requirements of the areas where those measures will later have to be put into practice. As in the Oporto Historic District, in the downtown of Coimbra, where video surveillance has been in operation since December 2009, this technology is looked on as 'a double-edge sword', in the words of its chief of police. He continued as follows:

> Firstly . . . it represents a burden for me, as far as the personnel I must release for that mission. The number of officers in this police station, as in all police stations, is always scarce. I understand that it might contribute to diminishing some feeling of insecurity that people may experience in Coimbra. Robberies in the downtown are no more frequent than in other areas; the city is not even targeted by large-scale robberies. But once one occurs, people run to the media, and soon a general feeling of insecurity sets in, even though there is no real insecurity in the sense that there is no relevant criminality in this part of the city.
>
> I don't know if video surveillance will have any practical effect, given that the number of robberies is not significant, but I concede that it may have some sort of effect on the population, deriving from the knowledge that these devices are being monitored by the police, keeping an eye on everything, so to speak.

Developing this argument further, the same officer explained to me what can be considered as the overall position of the Public Security Police regarding what clearly is a political initiative:

> We, the police, were in a way being caught in the crossfire. The law obliges us to collaborate; it was not our idea. I didn't feel the need for any kind of innovative device, let alone this particular one. I would prefer to have more personnel to assure a more intense surveillance, for the sole reason that the officer's presence is more dissuasive of crime and comforting to the passer-by: they know there is someone there.

One of the officers I spoke with in another part of the country stressed that although the Public Security Police is called on to participate actively in the process of putting together projects – through an assessment of crime rate statistics in the targeted areas – they were not consulted as to the means considered by them potentially more useful for more effective policing, a claim which supports the idea that projects follow predetermined conditions rather than constitute truly collaborative processes. On this point, James Scott's reflections on dominance and resistance are relevant when he states that:

> The theatrical imperatives that normally prevail in situations of domination produce a public transcript in close conformity with how the dominant group would wish to have things appear. The dominant never control the stage absolutely, but their wishes normally prevail. In the short run, it is in the interest of the subordinate to produce a more or less credible performance, speaking the lines and making the gestures he knows are expected of him. (Scott 1990: 4)

A similar rationalization of these arguments was repeatedly conveyed to me by police officers, who claimed that the idea to install video surveillance comes from higher up, and that therefore they complied with it dutifully, but could not acknowledge that it was essential, even in terms of material equipment, for the services they were called to render as the entity responsible for ensuring public law and order. If there were to be any funds to spend on material or logistic resources, they should be spent on human resources – whether to hire more police officers to support local precincts or to incorporate other teams, such as the crime investigation unit – and equipment needed for patrol cars, or even infrastructures such as police stations.

Despite the explanations offered to me in the course of the fieldwork, these officers recognized that their attitude was in total contradiction with the official position assumed by Public Security Police's national director, as well as with his public interventions, official communications or any documents that I might have had access to. Backstage manoeuvres played just as important a role as anything that was publicly announced, since it was there that decisions were taken on many issues, namely regarding concerns with security, evaluation of police activity, analysis of crime rates etc.

In Lisbon, the Public Security Police's Central Office offered even more criticism. After several meetings with the Ministry of Internal Affairs, during which the officers attempted to explain that a video surveillance system was not essential to them, and after being repeatedly given explicit orders to cooperate, the Central Office received the news of the Data Protection Authority's refusal on two occasions to introduce video surveillance in the Baixa Pombalina in Lisbon with some satisfaction, since it seemed to confirm their own scepticism regarding the use of the system in that part of the city. Despite the benefits that investment in cameras might bring, on the whole, police forces are reluctant to call them a first necessity good, claiming instead that the funds should be allocated to other equipment such as geo-tracking systems for patrol cars, intercommunication systems or more sophisticated databases.

This reluctance is connected not only to police forces' strict assessment of this system's efficiency and requirements, but also to the innate resistance of an organization that feels its capacity and ability are being questioned, as Gledhill suggests in his analysis of Foucault: 'Since programmes of power elaborated in discourses must be implemented through technologies that encounter the recalcitrant material of real societies and real people, their practical effects are determined by strategies (of domination and resistance)' (Gledhill 2000: 150). For the police and the policemen, pro-active intervention is considered a priority: police officers must be out in the street, ready to intervene directly when the occasion arises. Operating a monitoring device implies redirecting manpower from those other duties, thus weakening the more important human factor. According to them, the financial cost of this initiative does not compensate the material and human investment that is required for it to have a minimum efficiency. Even if we believe that the camera is the equivalent of, or may substitute, twenty pairs of eyes, in practice it results in removing agents from the far more essential 'action in the field', as they themselves call it, a fact which becomes all the more grave since there will be no agents available to substitute them out there.

This technological device, the video surveillance camera, while proving to be advantageous mainly in a post-crime situation, symbolizes the end of what has been termed *patrulha de proximidade* (proximity policing)[8] – that is, police intervention in the field, with an emphasis on direct contact with the population it protects. According to every officer to whom I spoke – whether they were for or against video surveillance – proximity policing is considered the most efficient and productive means to guarantee citizens' security and safety. These officers claim that proximity policing within the communities will be seriously affected, and that the failure to accompany this technological modernization with the strengthening of police contingents will ultimately render their action less effective.

The different police force representatives with whom I came into contact shared the opinion that the presence of surveillance cameras might only have the effect of diverting criminal activity to other unmonitored areas. Therefore, while it is important to ponder the equilibrium between the extended gaze provided by surveillance cameras – as an instrument that maximizes police resources and promotes swift and focused police intervention – and the diminished uniformed police presence on the field, it is just as essential to consider the tension between these systems' preventive and dissuasive functions, and the possible side effect of spreading criminal activity to unmonitored areas. Nevertheless, police officials were unanimous in recognizing that recorded images provided by video surveillance are unsurpassable in terms of their value as material evidence in crime investigations and criminal prosecution. Besides allowing for positive identification of criminals and criminal action, surveillance systems are also valuable tools as police training and research devices, allowing operational units 'to identify characteristic modus operandi of criminal groups, or study body language and facial expression even when individuals appear concealed behind hoods', as one officer stated.

The following case is a clear example of what William Webster has classified as the myths generated around video surveillance regarding the virtues of this monitoring system: that 'CCTV Works'; 'CCTV is everywhere'; 'citizens desire CCTV'; 'citizens understand the technological capabilities of CCTV'; and 'CCTV is there to protect us and to reduce crime' (Webster 2009: 17–19). This account, given to me by a member of the Public Security Police, relates to a street in Coimbra which is predominantly a commercial and recreational area during the daytime, but at night turns into a drug-trafficking spot. It amply demonstrates the cameras' purely dissuasive function, for although the cameras – conspicuously installed – were not yet in operation, their effect on the criminal activity registered during that short period was nonetheless significant:

> I am going to give you an example of a case where surveillance cameras dissuaded crime. There was a drug trafficking area here, where CCTV had recently been installed. Shortly after, there was an operation in which the Public Security Police's Crime Unit arrested three or four individuals for drug trafficking in the area. The cameras were not functioning at the time although already visibly in place.
>
> As soon as word got out on the street that the arrest had taken place, everybody – including the group that had been arrested – assumed it had been due to the action of [installing] the cameras, and I assure you that we won't be seeing any more drug trafficking-related arrests there in the time to come. Still, it is a risk to shift this sort of activity around; we know there are people who live exclusively on what they make from crime, and they will certainly continue to do so somewhere else.

The majority of people are not always aware of the location of existing cameras, and even when they are, cannot be sure if they are operating at any given moment, nor if (or for that matter, by whom) they are being monitored. My point is that the reliance on the dissuasive action of cameras – on its mere presence in given area – entails the assumption that subjects are aware of this monitoring mechanism as well as its functioning and capability. But even where that may be the case, I still believe that trusting unreservedly in the power of surveillance cameras to influence citizens' behaviour is a mistake, insofar as whatever the merit this system may possess on that score, it is to a certain extent annulled by the detached relationship that characterizes the multiple instances of interaction produced by such systems. I am thinking not only of the primary relation between the observer and the person under surveillance, but also its more subtle intermediate variations: between the observer and the visual document; and between the person under surveillance and the monitoring device.

Surveillance cameras are physically distant objects conceived to offer resident or transient populations in a given geographical area the promise of added safety. Nevertheless, according to the understanding of the police, the presence of cameras alone is simply not enough to ensure it. On the contrary, they believe that face-to-face interaction between security forces and citizens, the presence of an authority figure that represents order – ensuring social control and the correct deployment of its mechanisms – is irreplaceable as the most effective means to give real substance to that promise of safety.

Political Forces and Party Strategies

> Ladies and gentlemen,
>
> In modern societies, security – a fundamental right of citizens and at the same time an essential obligation of the state – is a required parameter for assessing the quality of democratic life. It makes no sense to talk about modernization, social and economic development, rights, freedoms and guarantees or the full exercise of citizenship, forgetting that the future is only possible with security for all.
>
> The democratic society, open and global, in which we live is, by nature and definition, a risk society. Dangerous situations acquire new dimensions. A framework for the emerging threats requires an equally innovative response strategy. In the area of Internal Affairs, the Government is developing a work that gives us the tranquility of knowing that we are performing our duty. However, we always want to do more and better.
>
> We value the commitment made to the Portuguese and we are determined to accomplish the goals we want to achieve. Internal security, public peace and crime prevention are tasks that the Ministry of Internal Affairs considers a high priority. Eradicating the sources of insecurity, preventing crime and prosecuting the perpetrators are tasks that cannot be delayed – tasks of the community and for the community, that concern and relate to all, from which all benefit and that require an integrated approach.
>
> Our strategy honours the Government's programme and embraces the idea of a collective security achieved through proximity policing, special youth protection programmes and similar initiatives for other particularly vulnerable groups such as the elderly. It also promotes systematic action to contain and control sources of risk, taking full advantage of emerging and innovative technology.

This declaration was presented by the Minister of Internal Affairs Rui Pereira in the symposium Safe Portugal Strategy (*Estratégia Portugal Seguro*) in 2007, roughly coinciding with the public announcement of the National Video Surveillance Programme. It clearly reaffirms the paradigm of security viewed as a collective and universal right of all citizens that it is the responsibility of the state to ensure, closely following an ideological line traced by Michalis Lianos:

> Post-industrial societies largely experience freedom as competitive individual participation in work and consumption. In many ways, security inevitably becomes the most indispensable part of freedom when freedom is exerted via capitalist competition. Surveillance presents itself then as a precondition rather than an antagonist of democracy, [a] necessary precaution for governing a society where individuals combine available ready-made options in their effort to distinguish themselves from others. (Lianos 2010: 69)

Among the discourses and practices of my informants, a common feature is their shared belief in this paradigm of security, which implies that the state is responsible for activating the mechanisms that ensure public order and safety, and for the achievement of this goal may sometimes require the parameters of personal freedoms to be reconfigured. If the discourses of members of the different political forces are analysed in light of the dichotomy contained in the above-mentioned principle – that is, collective security versus individual freedom – two major positions can be clearly distinguished. The Centre-Right parties clearly favour video surveillance. The Partido Social Democrata (Social Democratic Party, PSD) and Centro Democrático Social (Social Democratic Centre, CDS) both demand that the government, which they publicly accuse of incompetence, exert more efficient control over and punishment of criminals through measures that include installing video surveillance systems, strengthening police forces by increasing their manpower and updating their equipment, and applying harder sentences to offenders. Left-wing parties, such as Bloco de Esquerda (Left Bloc) and the Partido Comunista (Portuguese Communist Party), strongly oppose the use of surveillance cameras in open streets, considering that it constitutes an assault on freedom. Instead, they contend defend that policies should act preventively in anticipation of the phenomenon of criminality, focusing on its causes, such as unemployment or poverty.

Although these positions stem from political parties' different ideological traditions, one should be cautious about interpreting them too categorically. Most of the statements I obtained in interviews with politicians were ambiguous, if not ambivalent, especially when the project for video surveillance implementation in public areas was in its initial stage. I found, for instance, that some members of the Centre-Right were suspicious of the effectiveness of video surveillance, and expressed their concern over its repercussions on the right to privacy and free circulation. According to them, it was essential that the National Video Surveillance Programme's progress should be closely supervised. Conversely, there were also left-wing politicians who believed that the insecurity (whether statistical or psychological) felt mainly by urban populations had to be confronted, and admitted that this implied restructuring police forces and using all technological devices available.

To illustrate how party logic and the positions assumed by the different political groups operate throughout this process, it may be relevant to give a brief account of the parliamentary debates between 2005 and 2010 regarding public video surveillance, and more specifically, the discussion surrounding the enforcement of Law No. 1/2005. It should first be pointed out that this discussion was essentially instigated by members of the CDS, who on two different occasions made motions to present resolution proposals on this law. The first proposal, resolution 234/x, was proposed on 3 December 2007, and moved to create a National Video Surveillance Programme, which would actually be put into practice, albeit not in the terms proposed by this party. In this document, the CDS–PP 'urges the government to approve a plan of action to develop and install video surveillance throughout the country, as well as inform the population' (http://www.en.parlamento.pt/). A few years later, in 2010, they suggested a slightly different version of the previous proposal, which they called the National Video Protection Programme. It was inspired by the French model, which had been successfully implemented in that country between 2008 and 2010, partly due to a public campaign that had replaced the term 'video surveillance' by others with a less invasive connotation, such as 'video protection', 'video prevention' and 'video tranquillity'.

Following Bruno Latour's argument in *The Making of Law* (2010: 12) on the importance of transcribing passages from Council of State sessions for their value as 'textual matters', parliamentary sessions and their transcription are also considered essential in this work to understand the dimension assumed by the topic of video surveillance during its discussion within the chamber (see also Shore and Wright 2011). I must, however, make a parenthesis here to point out that, in contrast to the relative ease with which I gained access individually to members of Parliament who had taken part in these sessions, the shortage of actual occasions on which discussions on the topic had actually taken place in Parliament made it much harder to witness the debate in a more dynamic context, as one can do for example in studies where it is possible to organize with interlocutors as many focus groups as are found necessary. In any case, the two occasions that I did witness can still reveal something of the importance and impact that the theme of public video surveillance had on parliamentary discussion.

As mentioned above, both in 2007 and in 2010 this debate was introduced in Parliament by the CDS–PP. From the very start, this party clearly assumed its defence of a harsher penal system and its support for the use of more advanced and powerful crime-fighting technology, in which video surveillance emerged as an obvious candidate. During that period, the resolution projects and proposals that were presented in Parliament by the CDS–PP reflected this will to take on a more forceful approach. These instruments have been conceived merely to reinforce and extend the scope of the laws which have already been passed in Parliament. Resolution proposals are carried out following a standard procedure that usually takes place at the end of an ordinary session, after the main agenda for the day has been dealt with.

The resolution project of 2007 was met with strong criticism, namely from the more radical left-wing parties (Portuguese Communist Party, Left Bloc and Greens). But sceptical voices were also raised within the Socialist Party itself, despite having been the creator and first proponent of the National Video Surveillance Plan, which theoretically was consistent with the aims of the resolution projects. However, the Socialists' political opposition to the CDS–PP (and their intention to proclaim themselves as sole defenders of 'Portuguese safety'), and more importantly to the Social Democrats of the PSD (who joined the latter's side purely for strategic motives), proved more important than the actual debate of the measures at hand.

This resolution project was intended to formally bestow on Parliament the power to decide the places where public video surveillance should be installed, following the prior assessment and recommendations contained in a report submitted by police forces. At the time, the Ministry of Internal Affairs was still awaiting a decision on the authorization of the projects in Praia da Rocha and Oporto's Historical District. The delay in this final step was causing some frustration with the time it took for surveillance cameras to be effectively installed, especially considering that these projects had initially been hailed as a victory over those who opposed public video surveillance. This political initiative was therefore a reflection of the impatience felt by some members of Parliament with the whole process.

On 24 September 2010, the same political party presented a resolution project (262/xi/2) with a similar intention, but this time '[u]rging the government to design and approve the installation of video protection systems in areas of public use that are considered critical, within the scope of a National Video Protection Plan' (http://www.en.parlamento.pt/). In the proponents (CDS–PP)'s opinion, this measure would tackle the current ad hoc manner in which the existing plan was being enforced (given the small number of places where public video surveillance was operating), by assuming the decision to proceed with the systematic installation of video protection nationwide.

On that day, the main debate lasted approximately three hours (having started at 10:00 am), so CDS–PP only had a chance to begin their presentation and discussion of the proposal around 13:30 pm, after all other proceedings had been attended to. It is difficult to say whether what happened next was due to the lateness of the hour (already well into lunchtime) or to political indifference to this discussion, but the fact was that when the time came, most of the members present during the previous discussions had left, leaving the assembly down to almost one-third of its normal attendance. The Minister of Internal Affairs himself, present during the whole of the main debate, given that other security-related issues were also being voted on that day (a revision of the firearms law, for instance), abandoned the room with the rest of the government before the resolution proposal on public video surveillance was discussed.

The events that took place in those sessions of 2007 and 2010 perhaps reflected the lack of interest that this topic held for all other members of Parliament, but above all, they highlighted the performative nature of this kind of initiative. In fact, I do not think that this particular case was different from other, similar sessions that I witnessed, especially those involving initiatives that only require discussion but not voting, and in which member participation is more or less optional. Thus, proposals that had the ambitious intention of 'urging the government' were transformed into minor discussions with little parliamentary representation, closer to a mere formality than a serious political debate.

Even though the arguments were virtually identical in both debates – with the Communist Party and Left Bloc emerging as the opposing voice; the centre parties on both sides (Socialists to the Left and Social Democrats to the Right) maintaining an ambiguous stance; and the conservatives of the CDS–PP asserting themselves as its most determined defenders – one aspect had clearly begun to stand out, and that was the general reaction to the Data Protection Authority's position, particularly regarding the binding power of its decisions. As discussed above, the orientation that transpires from its rulings seems to be in total contradiction to Rui Pereira's idea that 'video surveillance is an instrument of freedom'. During this debate, some political leaders took their criticism of the Data Protection Authority even further, claiming that 'they are only concerned with protecting personal data, rather than protecting people!'

There is yet one important aspect worth considering regarding the coverage of video surveillance in open spaces in Portugal by the media and its presence in public debate. Its comparatively low media impact, reflected in the mutually reinforcing elements of its lack of popularity as a source for news and its failure to penetrate public debate and thus generate 'public opinion', can be summed up by a remark made to me by a member of Parliament: 'People are not electrified by the topic of video surveillance in public areas'. Nevertheless, the position taken by the Data Protection Authority, whose interventions began to be publicly problematic and conflicting, was also increasingly perceived by the general public as undermining a project that had supposedly been conceived to be swiftly applied across the whole country. Initially, the Data Protection Authority's decisions had not attracted much public reaction, but over the years they have resulted in some demonstrations of dissatisfaction, revealing a symptomatic lack of articulation and consensus among the various entities involved.[9]

Forgotten Diagnosis

This research followed the models of analysis used in political anthropology and in the anthropology of policy. This implies, on the one hand, a type of practical approach that includes preoccupations such as: working closely with interlocutors, recording their statements and following up on the consequences of those statements; keeping track of the various projects, their outcomes and resulting criticisms or conflicts; maintaining a general view of the whole process and a focused attention on each of the parties involved in it. Travelling to the different regions of the country allowed me to gain in situ experience of the different spaces, and a sense of their specific characteristics, that I could match with the arguments of their protagonists. On the other hand, at a more theoretical level, interpreting policies as 'cultural texts' became a particularly relevant tool, insofar as it often helped to shed light on certain issues that were hard to perceive merely through direct conversation or by visiting the places in question. On this point, Shore and Wright also stress the privileged role of the anthropological approach when they claim that 'anthropology's more open and democratic approach emphasizes not only the messiness and complexity of policy processes, but particularly the ambiguous and often contested manner in which policies are simultaneously enacted by different people and diverse situations' (Shore and Wright 2011: 8).

While the existence of conflicts, the difference of interests and even a certain amount of indifference to existing projects plainly transpired in conversations, the rulings of the Data Protection Authority, on the other hand, made evident the nature of those conflicts. Or rather, these rulings, as well as their corresponding dossiers (draft versions of the final ruling), contained all the official documents received by this organism for consultation, including all the steps, questions and analytical data, thus laying bare all the contradictions and objections that would ultimately be reflected in the final outcome. These files also told a story that was a complete version of what the different parties had verbally conveyed to me in fragmented and partial accounts. One of the points that this exercise made extremely obvious was precisely the lack of dialogue and communication among the different sides involved in this 'partnership'.

This miscommunication, however, is not merely a predicament of the institutions involved in this particular process. In my opinion it is indicative of a certain provincialism, constituting a deeper and broader national flaw. To use an example that may help to illustrate this point, I was surprised to realize that Lisbon's Public Security Police made little effort to exchange impressions with other local forces or visit sites where similar projects were being implemented by their counterparts in those areas. To me this was an obvious sign of regionalist competition and rivalry.

There is a relevant point to be made from this observation, and it once more links to Shore and Wright's comment that when policies are interpreted as cultural sources, it becomes easier to see that '[n]ot only do policies codify social norms and values, and articulate fundamental organizing principles of society, they also contain implicit (and sometimes explicit) models of society. . . . In many respects, therefore, policies encapsulate the entire history and culture of the society that generates them' (Shore and Wright 1997: 6).

This statement applies very accurately to the Portuguese context. Policy production and discussion in this country, regardless of whether one considers the authoritarian regime that ruled Portugal until the 1970s or its recent democratic history, have undeniably been characterized as being at the same time neutral and conservative. They are neutral in the sense that while there has been undeniably a tendency over the decades for relatively moderate political solutions (even during the so-called revolutionary period – PREC), on the other hand they frequently seem to lack conviction and consequence – almost as if their actual results were considered a minor issue. They are conservative insofar as after thirty years of an alternating democratic system, the image of dictatorship is still frequently invoked rhetorically to hamper the progress of new initiatives and ideas. I found an exemplary case of this spectral presence in the explanation given to me by an official of the Public Security Police in Coimbra, as to why the video surveillance project for that city's downtown had not originated in the police forces.Especially since my question did not focus on the necessity of this device or enquire after actual criminal data, but was posed as a merely theoretical hypothesis. The answer was definite:

> If it had been our idea, everyone would have immediately fallen on us with accusations that we were controlling, that we were using unnecessary force, and so on and so forth. Impossible. The police must always follow cautiously behind new developments; we never propose any measures of that kind because that will instantly make people say that we are returning to the time of the dictatorship!

The president of the Data Protection Authority had already pointed out this same argument during my first interviews with him. On one occasion, as he was telling me about the projects that had been ruled on and the processes leading up to the final decisions, he described how the council meetings were often marked by internal conflicts, assuring me that despite the final majority vote, decisions were far from consensual and always gave rise to heated debates. The president explained that one of the defining elements in these discussions was the age difference between the different council members. For instance, the youngest tended to contest the argument supporting the protection of citizens' image and privacy rights in recreational areas – as in the case of summer spots near the coast, characterized by their seasonal movement – claiming that in those cases the benefits of added security outweighed other considerations.

Inversely, older members, whose political memory stretched farther back and thus had a vivid recollection of the previous regime's political abuse, were prone to resist such reasoning and find it more objectionable. However, notwithstanding the opposing positions, so often distinctly expressed in those debates, the president would still claim that:

> Perhaps in this we are being more papist than the Pope. For example, when the Church asked to install surveillance cameras in Fátima, we were outraged. A religious site, a place of prayer, where people are supposed to be in touch with their innermost part? But it was the priests themselves who said that they had carried out a survey among their congregation, which showed that they were not at all bothered by the idea . . . . Still, we are very, very sceptical.

Let us return to James Ferguson, and the line of thought followed in *The Anti-politics Machine*, where the author demonstrates that even though the main goals of the Thaba-Tseka Project, in Lesotho, were to regenerate the region and promote its development through a substantial improvement of living conditions amongst its largely rural population, the results were not those expected. The project that was implemented in the city of Thaba-Tseka, like so many other campaigns initiated in Lesotho, apparently resulted in nothing. It did not achieve any of the goals proposed and was even pointed out as an example not to be followed (Fergusson 1994: 251–77). Nevertheless, as Ferguson reiterates, there is an alternative interpretation for this failure that can also serve to explain what might be considered a further, or implicit, project: that of the ramification and extension of control of state infrastructures in areas where they were previously absent.

That which Ferguson describes as the 'anti-politics machine' may be considered as one of the side effects of policy-making, and in itself is already a mode of political action, as in this case of an interventionist drive in Portugal. This work has already become a reference in development studies and its relevance acknowledged for a long time now. It is particularly useful here to show that the goals of development, modernization and improvement projects, regardless of the context in which they are implemented, easily become irrelevant or obsolete due to their failure to adjust to actual circumstances.

It may seem to be stating the obvious to point out that what works in one place may easily fail in another, but it is the underlying principle witnessed in cases like these that should give pause: that is, drawing a blank slate over the features of a specific context, based on the premise that a programme which has proven successful in one country of a given region or continent will necessarily have the same effect in another. Once again citing Ferguson on the African continent, we must realize that

> however diverse may be the empirical settings within which the 'development' apparatus operates, many aspects of 'development' interventions remain remarkably uniform and standardized from place to place. . . . But it is not only that 'development' interventions draw on a small and interlocking pool of personnel. More fundamental is the application in the most diverse empirical settings of a single, undifferentiated 'development' expertise. (Ferguson 1994: 258)

To do is to neglect or obliterate the historical, cultural, economic and political features of the given contexts, forcing an indistinct and undifferentiated model on them. On the other hand, the implementation of intervention models of this kind – based on the creation of infrastructures as a means to develop and modernize a region – may have the effect (negative or positive according to one's perspective) of allowing new patterns and data to emerge which provide the information needed to identify spheres of action that were previously harder to intervene in directly.[10]

The fitness of *The Anti-politics Machine* for an interpretation of the Portuguese context is particularly well adapted to the reading we can make of the National Video Surveillance Programme insofar as it allows us: (1) to contemplate a project supported by a government from beginning to end, giving one the chance to witness and record initial intentions and goals as well as accompany the development of subsequent stages of implementation and the transformations of the original plan; (2) to discuss the modes in which these projects are conceived and applied, focusing individually on each of the major parties involved; and (3) not to limit an appraisal to the end results – that is, simply to determine the degree of failure or success

– instead promoting reflection on the meaning and implications of those results. Michel Foucault's theories on discipline, governmentality and bio-power are on this account also relevant. My concern here is the conception, production and application of policies whose effects, despite surpassing and in some cases even contradicting initial expectations, are equally important for its proponents, even if the same does not apply to the populations that are the object of these policies.

In the few cases where the installation of video surveillance systems went ahead in Portugal during the period under analysis, the measure had negligible effects in terms of crime reduction and deterrence, in terms of both the quantity and quality of the crimes registered subsequently in those areas. However, instead of concluding that this constitutes a failure of the National Video Surveillance Programme, insofar as financial resources were squandered, expectations frustrated etc., perhaps we should rather highlight the fact that, as surveillance systems were being discussed, their effects and defects analysed and cameras bought and installed, policy was made, power games set in motion and economic and social issues brought to the foreground of public debate. Maybe the error is to presume that the project to install video surveillance is strictly bound up with crime-fighting policies, diverting attention from its broader implications, such as the general working of party-political dynamics, and other cultural, economic and social issues in Portugal.

Just as Ferguson's study of Lesotho explains that a standard model was applied to projects for different areas and regions on the (false) premise that they would accordingly have similar results, in Portugal it is clear that the implementation of imported models, in this case models of security based on public video surveillance, have been equally maladjusted. In both cases, this derives from the lack of a previous diagnosis of the local necessities of the places where such measures were to be applied, and from the absence of studies on the different elements and variables that might have significant bearing on those particular contexts.

At this point I would like to endorse the importance of James Scott's 'state simplifications': more specifically, the five features distinguished by him, since they are so thoroughly applicable in situations where this absence of diagnosis (or rather, the existence of simplistic diagnosis) leads to apparently frustrated, or in any case unexpected, results.

> [S]tate simplifications, the basic givens of modern statecraft, were, I began to realize, rather like abridged maps. . . . They were, moreover, not just maps. Rather, they were maps that, when allied to state power, would enable much of the reality they depicted to be remade. Thus a state cadastral map created to designate taxable property-holders does not merely describe a system of land tenure; it creates such a system through its ability to give its categories the force of law. (Scott 1998: 3)

Further on he adds: 'Legibility implies a viewer whose place is central and whose vision is synoptic. State simplifications . . . are designed to provide authorities with a schematic view of their society, a view not afforded to those without authority' (Scott 1998: 79). As Scott explains, state simplifications are no more than the amalgamation of otherwise dispersed information that is based on 'utilitarian facts' which have a solely 'official interest' (Scott 1998: 79–80). It is the connection of these variables that makes it possible to unify information, assembling it for specific purposes while at the same time taking it out of context. These aspects, whose underlying significance is explored throughout this book, demonstrate that the attempt to apply models of uniformity to populations, groups or territories is at the heart of the vicissitudes that have characterized public video surveillance projects in Portugal. The Portuguese context reveals that the contagious effect of European urban modernization and security models is allied to a populist and sensationalist attitude which is simultaneously political in all its infinite combinations. As will be discussed in Chapter 5, the feelings of insecurity attributed to populations are a form of anxiety that despite permeating the general discourse throughout major cities – a discourse that is as much a political statement as the product and producer of media information – finds no consubstantiation in the crime statistics contained in the Annual Internal Security Reports.

The argument suggested here is that in order to think about feelings of insecurity, which is not the same as thinking about security in the strict sense, one must focus on the multiple factors that contribute to their verbalization. If feelings of insecurity are to be granted any real existence, their analysis should not be based solely on issues of 'criminality', but also on other elements of transmission, such as common perceptions regarding the national economic and social situation at the various levels of society, returning to our initial indicators: health, education, employment, etc. Bearing these factors in mind, one cannot ignore the inherent subjectivity of social relations (Zurawski 2010, 2007). Fear, like other feelings, is as much an individual phenomenon as a collective one, depending on a wide array of other factors; the same can be said about feelings of insecurity. In the face of this subjectivity, the only element that can be truly said to determine the successful implementation of a project is its ability to 'improve' existing conditions – and the term is used here in its broadest possible sense: to improve education levels, standards of living, infrastructure etc.

Regardless of whether this impetus is actually materialized or remains in the sphere of intentions, 'improvement' remains the operative term here. In her work *The Will to Improve*, Tania Li, based on research conducted in Indonesia, states precisely that she wanted 'to understand the rationale of improvement schemes – what they seek to change, and the calculations they apply. I also seek to understand their effects, as they intersect with other processes shaping particular conjunctures' (Li 2007: 1). Thus, just like Ferguson, Li identifies several areas where those 'improvement schemes' suffer setbacks resulting from the frustration of their initial premises, going on to give a detailed and comprehensive historical account of the different interventions that the Indonesian archipelago (the author focuses mainly on the island of

Java) has been subjected to over the last couple of centuries, namely from the time it was a Dutch colony until its independence in the 1940s.

Following the works of Foucault, Marx and Gramsci, she identifies what she calls the two key practices required to turn this 'will to improve' into concrete programmes or schemes. The first is to acknowledge the existence of flaws that need correction in a given region. The second is what Li terms to 'render a set of processes technical',[11] meaning the whole process of gathering information needed to diagnose and decide the course of intervention to fight those deficiencies (2007: 7). It is this latter point that I wish to concentrate on, since it is the one that offers greatest room for interpreting the conception and implementation of projects, in line with Ferguson's observations. This practice, as Li explains, is largely based on the work of trustees and experts whose mission is to turn into a non-political issue the problem that being acted upon, as if it were possible to dissociate their diagnosis from the underlying political and economic intentions, making it a mere problem to resolve rather than something that can be used as a political weapon. 'For the most part, experts tasked with improvement exclude the structure of political-economic relations from their diagnoses and prescriptions. They focus more on the capacities of the poor than on the practices through which one social group impoverishes another' (Li 2007: 7).

This observation, closely resembling those of James Scott, is relevant for reflection on the nature and mission of similar programmes (whatever their sphere of action may be), as well as on their concrete forms and implementation. It allows this analysis to go even further, insofar as it highlights the discrepancies and modes of interaction among the different interlocutors in these projects, repeatedly evident regarding the National Video Surveillance Programme, which was characterized by inconsistencies and veiled antagonisms. The moment that the different experts start acting on a given problem, their first mission is to draw up a detailed plan that needs to be as broad and flexible as possible.

This task of emptying phenomena of their context carries with it other consequences, and the greater the number of actors involved, the greater the chance there is for misunderstanding and miscommunication throughout the process. The result is that as each institution follows its own political, cultural and economic agendas, so the diagnosis – originally intended to make sense of previously scattered data – becomes itself incoherent and unbalanced. The first elements to be discarded are the contextual specificities of the areas acted upon, following a logic eloquently described by Tania Li when she explains, for instance, that as she read the documents included in those improvement schemes, they confirmed that her investigations exposed 'multiple gaps: gaps between one document and the next, gaps between the world conveyed in the texts and the world to be transformed, and gaps between what the programs proposed and what they delivered' (Li 2007: 123).

In the Portuguese case this chain of events is reflected in how the different video surveillance projects set in motion in the period between 2007 and

2010 were marked by the repetition of the same mistakes and discordances, revealing a flawed learning process where information did not seem to have any cumulative effect, as if projects were making a fresh start each time. In my opinion, the apparent paradox of a wide-reaching modernization plan that is contradicted by the individual agendas of its constituent parts raises one major question that is hard to answer: what is the nature of the underlying forces that allow so many projects and plans all over the globe to be conceived and applied following a common set of premises – disaffiliation, decontextualization and uniformity of diagnosis carried out by experts whose practice of privileging one set of information over another ironically often has the effect of frustrating their own original goals?

Notes

1 See the Portuguese Data Protection Authority's website (http://www.cnpd.pt).
2 *Ibid.*
3 *Ibid.*
4 Dia Europeu da Protecção de Dados, Data Protection Authority, 28 January 2010, http://www.cnpd.pt/bin/relacoes/comunicados.htm (accessed 10 April 2013).
5 A good example to reflect upon the implications of disclosure of personal information on the Internet can be seen in its use by children and adolescents when participating in online games and virtual playgrounds such as the ones studied by Kerr and Steeves (2008).
6 Dia Europeu da Protecção de Dados, Data Protection Authority, 28 January 2010, http://www.cnpd.pt/bin/relacoes/comunicados.htm (accessed 10 April 2013).
7 As Susana Durão observes, 'At the same time that the state and the police present themselves as bureaucratic and administrative structures of control, they are constantly challenged both internally by their own officials and externally by the urban citizens. But it is on the former that the police and the State rely on to produce knowledge and police information, more specifically "crime rates". Even though held to be the safest source of information regarding criminal activity, every policeman knows that statistics are in practice easily manipulated by politicians who can overstate or understate them according to convenience' (Durão 2008: 18).
8 The main goal of the proximity policing model is to obtain the maximum possible integration of police forces within their designated communities, namely in terms of daily contact with populations. This is achieved, among other things, through the constant presence and visibility of agents, especially of teams on the beat, and the existence of neighbourhood police stations with six to eight police constables (Durão 2008, 2011).
9 The work of Steiner et al. (2004) is relevant on the subject of the analysis of parliamentary discourses.
10 For the case of Lesotho, Ferguson presents various reports from the World Bank (responsible for financing many interventions in this country), including an

exhaustive list of all the countries that contribute to these 'development projects', which includes the significant presence of some European countries, the United States of America, Saudi Arabia and Australia, among others. The author also lists the requirements a country must fulfill to become a potential candidate for these interventions.

11 She describes it thus: 'To render a set of processes technical and improvable an arena of intervention must be bounded, mapped, characterized, and documented; the relevant forces and relations must be identified; and a narrative must be devised connecting the proposed intervention to the problem it will solve' (Li 2007: 126).

Chapter 4

Public Issues, Private Matters

Thus far, the main parties involved in the process of video surveillance in Portugal have been identified, along with their strategies, which can be summed up as follows: political forces appeal directly to collective concerns, usually focusing on the rhetoric of security; the media exhaustively publicize major or minor events, amplifying their impact and notoriety. Furthermore, academia sounds an alarm about the dangers of the widespread use of surveillance devices, neglecting and even sometimes refusing to acknowledge that these political measures are in fact supported and even prompted by pressure exerted by citizens themselves. In the end, all political activity and concern with security is ultimately aimed at citizens.

When we confront issues of security or security devices, we must bear in mind that we are facing a theme that is susceptible to multiple interpretations. This is because the notion of security is not just about physical integrity but involves many other features besides the perception of risk or the feeling of fear in connection to the phenomenon of crime. It summons a broader spectrum of concerns that derive from socio-economic conditions, at the same time that it is constructed within the more intimate sphere of personal experience and sensibility, which are also culturally determined. As several authors have argued in their analysis of opinion polls regarding the use of video surveillance and its relevance to ensuring collective security, answers vary according to the perspective from which the questions are presented. The mere selection of a line of inquiry has an influence on the resulting responses. The same can be argued about the validity of studies that seek to identify the general trends and potential uses of surveillance and security systems, and their effects on the daily lives of citizens. In my opinion, opinion polls and barometers are only truly useful if the conclusions derived from data obtained in this manner result from a balanced examination of their virtues and limitations.

We must be particularly careful when analysing this type of material, remembering that, for example, even the studies on video surveillance which include an element of participant observation – thus including a qualitative methodology – are, more often than not, centred on the actions of those who operate the cameras (their experience and habits, how they determine what constitutes 'suspicious behaviour' etc.) rather than on those who are under surveillance.[1] We certainly learn a lot from both quantitative studies, based on large-scale surveys, and qualitative research on video surveillance monitoring operators, whose focus on the activity and behaviour of the watchers highlight relevant aspects of the electronic gaze. Notwithstanding these approaches, a holistic and comprehensive study of this phenomenon must also consider the perception that citizens actually have of this whole process, regardless of whether it is enforced as a result of political rhetoric, commercial opportunism or excessive securitarian zeal. An analysis that aims to provide an accurate and comprehensive study must be careful to avoid its own preconceptions, and to a certain extent even withhold judgment regarding citizens' perceptions of these phenomena.

In the next pages I will try to take my own advice while presenting my observations on Portuguese citizens' common perceptions of the major themes and issues discussed throughout the previous chapters. Focusing on the nuances of the Portuguese context, this chapter will now turn to the common citizens and their opinions on public video surveillance. This may help to overcome the deadlock that, in my opinion, some current surveillance studies have to some extent brought to this issue by insisting on putting the problem in terms of the distinction between controllers and victims of control. To put it in a slightly different way, my aim is to reflect upon the notion that there is a blind, overbearing and unified force exerting its power, and simultaneously, a dispersed, impotent and unaware set of citizens whose privacy and rights are constantly and purposely being intruded upon. However, it would be equally dangerous to lean towards the other extreme, and I certainly do not mean to dismiss the real risks and threats implied by surveillance mechanisms. It would be preposterous to think that such a powerful device is immune to excesses and misuses, or to ignore the fact that this field is prone to attracting interests that are not always totally transparent. I think that the exercise of putting aside the *a priori* conceptions which one might have regarding common citizens' reactions to video surveillance in public areas is one that is well worth attempting.

It is safe to say that in Portugal (though I suspect the same applies in other countries) the ordinary citizen is content with the idea that video surveillance is just a device used to catch criminals, and cameras a purely benign presence designed for their safety and well-being. Most people are not primarily concerned with other, less evident consequences. As will also be discussed in the following chapter, when questioned about video surveillance, the usual response is to think of its advantages rather than its perils, and the same is true regarding other related issues such as access to personal data by large corporations or their online disclosure. People show some concern when confronted with potentially negative consequences of such activities, but they are easily overpowered by the instinct to be compliant, whether because it is convenient for instance to join an online social network, or even attractive, to gain consumer benefits. The claim that video surveillance promotes minority discrimination or may be used as an instrument of voyeurism or extortion has a difficult time making its way into citizens' general awareness because these aspects are regarded as exceptional situations that do not affect the majority, even though this claim is actually substantiated.

However alarming or erroneous this common perception may seem to social scientists, this is an important element of the phenomenon, and one which may be useful for this study, even if that implies momentarily setting aside abstract theorization and moving into a more subjective and at the same time more empirical terrain. Some of the questions that will be discussed in this chapter may be considered simple and in fact, it is the simple and more common view that this section seeks to understand. What do citizens think of video surveillance? How do they respond when questioned about its advantages and disadvantages? What is citizens' perception of the role played by police forces as the guardians of public order and the use of surveillance cameras in Portugal? What are people's notions of security and liberty and how do they connect them to concepts like public and private spheres, or the relationship between individual and collective? These are all questions that have already been addressed in relation to the different organisms involved in the National Video Surveillance Programme, which leads me to wonder if the disparity between the agendas of the Ministry of Internal Affairs, Public Security Police and the Data Protection Authority somehow reflects onto the ordinary discourse of those who are targeted by these measures.

However concerned one might be about the present and future impact of surveillance technologies in social, economic and cultural life, it becomes obvious from looking at the discourse of those who are not directly implicated in these processes, being in a way its passive subjects, that the notions of security, privacy or freedom vary enormously, but are generally closer to the broader ontological conception of security as described by Giddens (1991). In terms of the significance of video surveillance, the emphasis in ordinary discourse is placed not so much on its actual effectiveness and suitableness, but on authorities' zealousness in ensuring collective well-being.

The qualitative data that served as the basis for my observations regarding the different perceptions of video surveillance was gathered from Portuguese national newspapers (Ponte 2004). I paid particular attention to readers' comments on news announcing projects of video surveillance in open spaces in different cities throughout the country. Out of the three daily national newspapers chosen, two are actually separate publications by the same editorial group, one for each of the two major regions of Portugal: the *Diário de Notícias* focuses more on news from the country's central and southern regions (including the capital, Lisbon), and its approximate circulation is around 65 000, while the *Jornal de Notícias* reaches a circulation of about 130 000 and focuses on Portugal's northern region, particularly on news that is of interest to the people of Oporto, Portugal's second largest city. The third newspaper, *Público* (circulation 60 000), also has special editions for each of these cities, and has a middle- and upper middle-class target audience (Nery 2004). This kind of analysis, based on material collected from sources such as mass media, forum debates, weblogs or other online channels of communication, has gained an increasingly prominent place in anthropological studies. As Gupta and Ferguson point out: 'Participant observation continues to be a major part of positioned anthropological methodologies, but it is ceasing to be fetishized; talking to and living with the members of a community are increasingly taking their place alongside reading newspapers, analyzing government documents, observing the activities of governing elites, and tracking the internal logics of transnational development agencies and corporations' (Gupta and Ferguson 1997: 37).

Broadly speaking, I have adopted Akhil Gupta's idea about the importance of 'public culture', which he defines as 'a zone of cultural debate conducted through the mass media, other mechanical modes of reproduction, and the visible practices of institutions such as the state' (Gupta 2006: 221). I believe that in this particular study, if I had chosen, for instance, to conduct interviews or directly speak to people from potentially relevant specific geographical areas, such as Lisbon, Coimbra or Oporto's downtown districts, I might have run the risk of quickly exhausting information that would perhaps even end up being journalistic and not very thorough. Even though the role of mass media in shaping public opinion (through the selection, control and treatment of information) is in itself a major theme for any study using this source material (Jewkes 2011), in this case the focus is onto the reactions of the audience to the information conveyed in these media than in the content of the news presented.

What immediately stands out is that between 2005 and 2010, news media repeatedly announced the imminent installation of video surveillance systems in open spaces in different parts of the country, which, it is now apparent, never actually occurred. Nevertheless, these news elicited strong reactions from readers and heated debate over the pros and cons of these systems regardless of whether they had been effectively installed. While the anthropologist seeks to give a distanced study, supported by the kind of inside information accessed in the course of fieldwork done backstage during these processes, the media frequently has a more circumstantial and immediate interest in the phenomena they cover. In this case, the media impact of certain issues and events, coupled with the criteria of journalistic treatment, results in information that does not actually correspond to reality because it ignores the bureaucratic and legal processes behind the scenes (Best 1989; Cook 2005; Curran and Seaton 2003; Franklin 1999). On the other hand, readers' reactions provide a more significant record of individuals' feelings at a given moment.

Besides interviews and conversations with residents and businesses from the areas affected by open space video surveillance systems (whether merely announced or effectively operating), the study of the news that explicitly mentions video surveillance in open spaces in these three daily newspapers allows us to identify the period in which there was greatest journalistic interest in the topic. This provided an overview of the period between 2005 and 2010, when the installation of surveillance devices in the streets of Portuguese cities received greater attention. The resulting picture traces approximately the same timeline of events as that described in the previous chapters, and is almost a copy of their development. Once again, 2008 and 2009 were the years in which the media most intensely concentrated on this phenomenon, maybe because a greater number of decisions from the Data Protection Authority regarding authorizations or rejections of projects were made public.

Considering the period between 1 January 2005 and 31 December 2010, *Diário de Notícias* registered 739 news articles under the category 'video surveillance', while *Jornal de Notícias* published 840 in the same period. This broad category includes all video surveillance-related events: nationally and internationally; both in open and in private spaces; including all video surveillance in state schools – as part of the Simplex programme, more specifically the Technological Plan; in other public institutions (hospitals, banks, courtrooms); and on public transport and roads. The number of news articles specifically on video surveillance in open spaces is much smaller, roughly representing 20 per cent of the total (170 for the central region and 199 for the northern region). This phenomenon only started to draw considerable media attention in 2008, even though the law that authorizes video surveillance in open spaces dates back to 2005. Some examples of sensationalism can be found in the headlines of newspapers during this period, such as: 'PSP [Public Security Police] Launches Big Brother' (*Jornal de Notícias*, 16 December 2009), for an article reporting that the video surveillance system in the streets of Coimbra was finally operating; or 'Video Surveillance Lingers while Vandalism Continues' (*Diário de Notícias*, 1 April 2008), in an article denouncing the delay in implementing video surveillance in Praia da Rocha, in the southern city of Portimão – a project which was never actually fully authorized on the terms requested.

While emphasis on news content accentuates the elements that determine how certain opinions are formed and communicated – whether by analysing a newspaper's editorial line and its preference for a certain type of subject matter over another, or by studying how journalists themselves work – I consider it more relevant for this study to concentrate on providing a detailed account of the reactions to that material, as it emerges in the discourse of newspaper readers, and especially in the written comments in the newspapers' online forums. This material contains a richness that cannot be found elsewhere, particularly in terms of its uncensored, unselfconscious nature. This allows the researcher to make an informal and undirected survey of emerging perceptions of the issue at hand, which is not defined by the researcher (as in an interview or a questionnaire), but arises from a piece of newspaper journalism that does not address readers directly about their opinions or beliefs. On the other hand, the comments' unprovoked nature diminishes the purely reactive element, so it may be ascertained that their authors have a spontaneous interest and opinion on these matters, and are not just reacting to enquiries.

A major source of information, as far as online comments from readers regarding news on this topic are concerned, is the website of the newspaper *Público*. Although these comments present several methodological limitations, such as the impossibility of confirming important data regarding their authors (origin, profession, gender, age), and of course the fact that this audience represents a very specific sample of the population (people who are in the habit of reading, and who have available time to spend on this activity and internet access), they still present enormous potential in terms of content and specific features. For instance, even though comments are submitted individually, in most cases they generate interaction between participants which closely resembles a forum debate or virtual chat room, where people discuss each other's ideas and offer critical judgement on the news in question.

It should be remembered that the virtual environment protects anonymity, allowing the use of pseudonyms, which is not always possible in other forms of written participation such as letters to the editor. This anonymous atmosphere encourages speech to become less constrained, or at least eliminates some obstacles to free speech encountered in direct social interaction. As the vast academic production on the uses of anonymity and pseudonyms in the virtual world has abundantly shown, and I explored theoretically and empirically in previous works (Frois 2009a; 2009b), when protected by the condition of anonymity, discourse and communication become free from 'normality' criteria. Political correctness is replaced by a freedom of expression that is hard to achieve in face-to-face interaction, especially when topics may involve, as in the cases presented here, comments involving racist, xenophobic or politically radical content.

Public and Private: A Matter of Opinion

The following example illustrates the kind of spontaneous reaction that was of interest in this research, regarding citizens' perceptions of these issues. On the occasion of European Data Protection Day, the radio station TSF[2] hosted an open debate on the theme 'New Technologies and their Interference in Private Life'. The Data Protection Authority's president, who was the main guest on the programme, took the opportunity to make the usual public declaration that has already been discussed in the previous chapter. He talked at length about the importance of raising public awareness regarding the protection of personal data and its role in ensuring citizens' freedom and security, and warning against the risks of ignoring this issue. The following are opinions expressed by some of the callers who participated in the debate that day:

> In public spaces what matters most is the safety of the passer-by, namely, older people and children who may be kidnapped or suffer abuse; and this is the importance of having video surveillance and other crime-fighting devices in public areas. Regarding the righteous people who freely walk through their local streets, I am sure they don't have any problem with being watched. As the saying goes, 'if you have nothing to hide, you have nothing to fear'.

Some key themes can immediately be identified in this particular excerpt: to begin with, the concern with everyday urban pedestrians, particularly those groups considered more vulnerable to crime. Above all, these comments express feelings of fear, and a profound belief in the security potentially provided by video surveillance. It may seem to be stating the obvious to say that fear is a common reaction to crime, but the feeling itself is a complex one deserving attention (Low 2004; Low and Lawrence-Zuñiga 2003; Low and Smith 2005). For the moment it is important to observe that the kind of fear implied in this passage is mostly a psychological one, especially considering the types of crime that are most commonly practised in Portugal, which, as will be discussed in the next chapter, can be classified as petty street criminality.

Nothing in this caller's testimony suggests that he has ever been a victim of the crimes alluded to: there appears no element of trauma to justify insecurity. Likewise, there is no indication that the caller has had any experience of the benefits provided by surveillance cameras. In the end, it is a case of blind faith in the efficiency of a technological device as protection against blind fear. This passage is included, not to dismiss this person's concerns, but precisely because it points to the crucial theme of objective versus subjective data in this kind of study. Another equally relevant argument implied by this passage, one to which this discussion will return several times in the course of the following pages, can be described as the belief that law-abiding, 'innocent' citizens have nothing to hide, and consequently have nothing to fear from being placed under surveillance.

This reasoning must be divided into its two underlying ideas. The first of these has to do with the notion of privacy and the role it plays in the life of the common, law-abiding citizen, which if this excerpt's reasoning were taken to its ultimate consequences, would be close to none. This person clearly states that he is prepared to trade any rights to image and privacy in public space, so zealously defended by the Data Protection Authority, for the peace of mind that video surveillance might bring to his daily routines. This gives the notion of privacy an almost negative connotation, as if it were only useful as a cover-up for wrongdoing. It is reasonable to conclude that the idea of privacy can sometimes be associated with the idea of crime, and therefore that the desire for privacy can be seen as suspicious.

The second, more subtle idea implied in this reasoning is associated with a phenomenon that will also be discussed further elsewhere, which is that a given society can be separated into two easily definable and immutable groups: the innocent and the guilty. It should be emphasized once again that the information interpreted here was not gathered from direct questioning, as would occur in a semi-directed interview or survey, but is a record of the undirected and spontaneous participation of radio listeners invited to join a public debate. Let us turn to another of these:

> Right now we have to choose between our right to freedom and our right to security in general. We live in a time when all of us are feeling extremely vulnerable. There are no conditions, not even legislative ones, for the common citizen to live well and feel comfortable walking in the street at any time of day, let alone at night. I don't know why people don't accept CCTV if frontiers between public and private spheres no longer exist anyway.

This second comment was among dozens that were broadcasted during the two hours the debate lasted. It follows a line of reasoning, concerning the right to freedom and security, which is close to the Ministry of Internal Affairs' official position as seen in the previous chapter, in the sense that it equates safety with freedom. These two concepts are often viewed as opposites or as constituting a relationship of inverse proportionality. Here, on the contrary, they are considered mutually reinforcing. Propositions such as 'For freedom to exist there must be security' or 'Security is a requirement of freedom' have already been mentioned, and the next chapter will deal extensively with these notions. Given the prominence of these expressions in everyday discourse, their meaning will now be discussed in greater detail.

First of all, it is evident that there are other aspects to consider in this context besides the notion of privacy. For instance, there are different perceptions and conceptions of public space. Sætan, Lomell and Wiecek suggest, in their study on public CCTV in Norway and Denmark, that

> a discussion of public spaces in terms of inclusion versus exclusion should not preclude the continued discussion of privacy versus protection. In fact, we see the two discourses as related on at least two counts. The safety and perceived safety of public spaces is an issue not only at the personal level (for victims of crime) but also at the social level, since spaces that are unsafe or are perceived as such will have reduced value as economic, cultural, and social capital. (Sætan, Lomell and Wiecek 2004: 399)

While concepts such as privacy, intimacy, public versus private space, freedom and security are widely used, they can be misinterpreted and become ambiguous. For instance, laws – conceived as abstract and general rules to regulate society and protect the rights of all citizens – are misconstrued by those same citizens for whom they are intended and who do not relate to such abstractions and general principles. As will be observed, the fact that nowadays there is much more information about individual rights and freedoms, and that there is a huge proliferation of mechanisms ensuring them, does not necessarily mean that people are actually more knowledgeable about such matters or that they are more aware of the consequences and risks to their rights and freedoms of the actions they take. The bottom line is that people generally do not perceive as a threat anything that does not affect them directly.[3]

This comment expresses its author's preoccupation with the increased feeling of insecurity, manifested in the fear some people have of walking alone in the streets at night. It further expresses his opinion that current laws do not ensure citizens' well-being, and that the justice system in particular is generally ineffective in pursuing and punishing felonies. There is another issue that emerges from this passage, whose significance is attested by its recurrence in the discourse of police forces and decision-makers, and thus can be said to form a shared conception. It has to do with the wider notion of public and private spheres and its applicability to specific sites. This caller's testimony questions the existence of a frontier between these two spheres, or at least claims that it has been blurred, and therefore raises doubts as to whether we can rightfully speak of privacy in the public sphere, questioning even the notion of privacy itself. Many of the statements that will be examined here claim that it is no longer possible to remain anonymous or inconspicuous anywhere, since 'surveillance' is present in all daily activities.

There is no simple answer to these intricate questions, and they will have to be approached gradually throughout, as they unfold. The previous excerpt also raises the question of whether the ideas of freedom and security and public *vs* private spaces have different meanings according to the circumstance in which they emerge. The obvious and almost immediate answer would be, 'Yes, these notions vary when applied to different persons with distinct social statuses and responsibilities'. If this is assumed to be true, the question must be asked, 'What in fact is the specific meaning that video surveillance in open spaces has for common citizens, those who are effectively targeted by the measures which allegedly they have no power to decide?' How do citizens evaluate the potential consequences implied by the use of open-space surveillance devices? The next pages will continue to give voice to the common citizen in an attempt to provide clues to the answers to these questions.

Video Surveillance: Security and its Nuances

In addition to what the comments presented below may reveal in terms of the subject matter, and which generally provides a personal view on issues related to the security of persons and goods, privacy, security and freedom rights, they also contain an element of carelessness regarding the effects they may have on others and even the consequences they may have for their own authors. It is frequent to find in these comments harsh and open criticism of politicians or public institutions and their agents, such as the police, the Ministry of Internal Affairs or the Data Protection Authority.

In addition, it is apparent that readers/commenters intervene precisely when they sense the impending effects of these measures on their daily lives, as for instance in cases reporting that video surveillance systems are about to be installed in their area of residence or business. This approach to journalism has already been followed by Bird (2010), who explains that the comments on an article may contain additional and significant information for the anthropologist's study of certain phenomena. 'In an analysis of online discussion of a story . . . the story was used not as a "text" with a clear meaning but as an opportunity to interrogate issues, from morality, to religion, to race. People do not evaluate news stories in isolation but incorporate them into their already established worldviews' (Bird 2010: 14).

Generally speaking, these comments express opinions, but more importantly those opinions convey very specific feelings. When writing and commenting, readers are not merely focusing on the article's themes, but on their own reactions to those themes: agreement and support or suspicion and resistance, to name just a few. The overall tone of discourse can also be revealing, and the combination of this bundle of information with the irony, anxiety or derision that often transpire in comments results in a rich and multifaceted portrait of public perception of the idea of installing surveillance cameras in open areas. The following comments illustrate this. They were submitted by readers of the *Público* newspaper when authorization for Lisbon's Bairro Alto video surveillance project was announced, on 26 November 2009.

> *Please, by all means film away*
> I will never understand people's problems with surveillance cameras. As far as I'm concerned, as long as they don't come into my house it's fine by me; innocent people have nothing to fear. Any measures to help track down and catch criminals are welcome.

As in the comment in support of video surveillance implementation mentioned previously, the aspect that stands out here is the argument that privacy is a non-issue when dealing with the public space. This comment emphasizes the idea that the sphere of privacy is absolutely restricted to the space of the household, and therefore as long as this is respected, as long as 'they don't come into my house', video surveillance does not in any way jeopardize it.

Another relevant issue is also addressed here, when the commenter refers to 'people's problems', implying that there is a general underlying objection to being watched by others, while it expounds the potential benefits that would come from this practice. This statement shows belief in video surveillance's effectiveness as a crime-fighting instrument, to 'track down and catch criminals'. The next excerpt touches upon the same issues, but in a more elaborate fashion, weighing up the pros and cons of the use of surveillance cameras and even questioning its effectiveness and implications in terms of these devices' invasion of privacy.

> With so many conditions [referring to the legal restrictions to images that capture doorways and windows, and the prohibition of facial recognition by police forces,] I'm afraid cameras may become useless, especially given the large number of blind

> spots. But this question of rights and privacy, although extremely relevant, has increasingly become a non-issue, given that nowadays every little corner shop owns CCTV equipment, let alone supermarkets, banks, or even shopping malls, which are nearly public spaces. In all these places, citizens willingly accept being monitored, at least partially renouncing some rights in exchange for a supposed security.
>
> Of course no one is forced to enter such places, but the relinquishing of privacy is not limited to being captured on film, extending to the use of credit cards, customer loyalty cards, mobile phone communication, etc. All these situations are liable to surveillance, and the fact is that in most cases they are to a greater or lesser degree effectively monitored by private entities. I consider that video surveillance is, however unfortunately, necessary in some areas, and therefore I obviously accept giving up the right to privacy if that means improving my safety or serving to dissuade criminal activity such as drug trafficking.

The line of thought presented here, especially its admission that video surveillance is already widely disseminated in public spaces, can be compared to the description in the last chapter of political debates in Parliament, where the CDS–PP party tabled a motion expressing the need for a National Video Protection Plan. It is indeed a relevant point to consider, especially in terms of discussing video surveillance no longer just as a potentially useful device, but as an inevitable part of life. Following this reasoning of ubiquity and inevitability, there are two alternative responses that can be summed up in the questions, 'If video surveillance is already everywhere, what further harm can it do in public space?' or inversely, 'If it is already everywhere, what further good can it do in public space?'

The contexts alluded to in this comment also bring to mind concerns expressed by the Data Protection Authority, insofar as the commenter acknowledges and identifies other forms of monitoring and controlling citizens for a variety of ends (commercial, financial, legal etc.). Nevertheless, the primary argument, contained in this commenter's closing remark, has to do with accepting 'giving up the right to privacy' in favour of one's own and others' strengthened security. In short, it shows that the preservation of security, not only as a personal concern but also in terms of society as a whole, prevails over other rights. The next chapter will return to this conception of security, but for now it is enough to stress its pervasiveness throughout the comments under analysis, and to note that it also frequently appears under more political guises, when it is equated with the notion of freedom, as in arguments such as 'there is no freedom without security' or 'security is a right that ensures freedom' that were expressed to me during interviews with members of Parliament.

On the other hand, those readers who oppose video surveillance often focus on the dangers of this security argument, warning against its excessive use and considering it a first step in a process that inevitably leads to thwarting the freedom of citizens. The following comment considers security-enhancing measures and their possibly negative effects, by reviving what may be called the ghosts of the past.

> *Big Brother*
> The danger in this kind of thing is to open the door . . . Then, when later the Parliament is set on fire, the communists are blamed (or the Muslims, whatever is fashionable at the time), [and] one day we wake up to find that we are ruled by a man with a ridiculous little moustache who believes that he OWNS THE TRUTH.[4]

This commenter's metaphorical reference to Adolf Hitler, to political, religious or ethnic persecutions and to George Orwell's Big Brother show that video surveillance can also be perceived as a threat that can lead to dangerous uses such as discrimination, wrongful use of information and even segregation of groups within society. The implication here is that more important than discussing the video surveillance per se, its features and effectiveness, and weighing these against the abstract notions of civil rights and freedoms, is to consider the uses or in this case the misuses of such measures. There is clearly a shift of perspective in this approach which brings a new element to the equation, the element of intention, which is closely connected to notions of power, especially political power, that are extremely relevant in the Portuguese context. The experience of a dictatorial regime in Portugal, of a history of oppression, is easily summoned even by people who never actually lived it. Note that in the composite picture imagined by this commenter, it is still communism, the movement that was most strongly persecuted during Salazar's regime, that is singled out as the obvious scapegoat. Another relevant point made by this comment is the acknowledgement that the targets of video surveillance, or even surveillance in a broader sense, for that matter, can change according to historical and political context. The commenter's reference to Muslims or 'whatever is fashionable at the time' also questions the notion of suspicion and its changing nature, a category that gained very specific features in the post–September 11 context, after the attacks on the World Trade Center and the Pentagon (Aoude 2004; Rowlands 2004).

What these comments reflect about the belief in the efficiency and effects of video surveillance has already been addressed by other authors studying the spread of video surveillance in urban city centres, who explain that that it is 'the testimony of belief in CCTV as a crime fighting device that contributes to increasing safety' (Webster 2009), and that generally, concerns with the violation of privacy and other rights are overridden by conviction in the popular wisdom contained in dictums such as 'the innocent have nothing to fear', or 'if you have nothing to hide, you have nothing to fear' (Bannister, Fyfe and Kearns 1998: 21 and ff.).

Privacy: A Right for Everyone?

As mentioned above, many readers' comments take an ironic or even sarcastic tone when addressing both the actions of politicians – often accused of incompetence and idleness – and the defence of privacy as an argument to prevent the possible installation of surveillance cameras. As is evident, this topic is a recurrent theme. But although the question of privacy is frequently addressed in comments, it is always in very broad and even vague terms, rarely reflecting on any of its deeper implications or discussing its concrete forms, such as image rights or the right to free circulation. I therefore believe that this recurrence does not reflect any special interest in issues of privacy on behalf of citizens, but is merely a reaction to a topic that is very often the focus of the journalistic pieces being commented on.

Whenever privacy is mentioned, people immediately associate this concept with a very restricted space: the home, the room, the bed, the bathroom etc. It might be said that privacy is primarily associated with the sphere of intimacy; with the protection of sentimental and family life, and the spaces they are commonly assigned to – symbolized by the home – and not as much with outdoor activities. The following article was published in the *Público* newspaper on 10 August 2010, at the time that a ruling was announced refusing authorization to install video surveillance cameras in Amadora district (one of greater Lisbon's most notoriously problematic areas in terms of crime). The argument used for this decision was that the existing crime rates did not justify subjecting the thousands of persons residing in the area to such monitoring.[5]

> *Ministry of Internal Affairs rejects video surveillance in Amadora*
> The Ministry of Internal Affairs did not grant authorization for the installation of a video surveillance system requested by Amadora council, on the grounds that it threatens citizens' privacy . . . The Ministry has 'doubts about this project's envisaged results and efficiency', emphasizing that the installation of such a project, 'given its extension and permanent character, would deeply affect the population, be particularly intrusive on the rights of individuals monitored by the cameras, and therefore threaten their fundamental right to privacy'.

This news article is particularly interesting as it reveals how journalistic technique and style alone condition not only the tone of the information conveyed, but even its facts – in this case the alleged rejection by the Ministry of Internal Affairs of Amadora's request for video surveillance. The information conveyed does not correspond to what actually occurred, and it even distorts the real motives behind this outcome. As has been observed previously, these processes follow a complex and strict institutional procedure, in which the Ministry of Internal Affairs is responsible for making an initial assessment of the project submitted by local councils, which, if positive, means the project can then go on to the Data Protection Authority to be reviewed and approved.

Thus, it is the latter institution that ultimately has the power to enforce decisions, after which the project can then return to the Ministry of Internal Affairs to be promulgated and the word given to initiate operations. The ministry is therefore dependent on the Data Protection Authority's decision, regardless of whether they agree with it or not. What happened in this case was that the Ministry of Internal Affairs supported the Amadora council's project from the very beginning and agreed with the arguments for its relevance and suitability. However, when this entity was faced with a negative response from the Data Protection Authority, it was forced to make a statement that apparently contradicted the position previously defended within the ministry. The news piece that was published, and transcribed above, is in fact a description of the ruling given by the Data Protection Authority, and not a faithful portrayal of the ministry's position regarding this particular case. But more than a case of deliberate misinformation, this piece above all reflects a misunderstanding that directly results from the kind of vicissitude already described in the previous chapter, and which characterizes the National Video Surveillance Programme as a whole.

The apparent arbitrariness and ambiguity of arguments, as well as the lack of transparency or coherence reflected in the content of this news article, ultimately do show that mass media can easily be led to give out partial, if not erroneous information. On the other hand, it is clear from readers' reactions to this piece of news and to the decision reported in it, that most people consider that the defence of privacy (both in the public and private sphere) and the protection of the right to one's image, free circulation and intimacy do not apply in this context. This is because they consider that these protections only benefit criminals and not law-abiding citizens. The lack of correlation experienced in everyday life between what citizens hold to be relevant to them and what is dictated by law is also responsible for making the latter so susceptible to widely varying interpretations, as is evident in some of the comments that ensued online, demonstrating the controversy caused by this text:

> *Thug's right to privacy*
> All criminals in Amadora and in all Portugal have the right to mug, murder, rape, deal drugs, etc. without being inconvenienced, it's their privacy that's being threatened, not that of tax-paying citizens who just want to move about safely.

> *Pseudo-freedom!*
> The Constitution also establishes the right to dignified dwelling, but Amadora still has many slums and degraded buildings. What do these pseudo-liberals intend? For women to continue being raped, Portuguese youths being mugged, and terror continuing to dominate the Sintra area? From the comfort of magnificent salaries that increasingly dissociate them from the populace, those public servants known as politicians have the luxury of acting as defenders of the Constitution . . . The people demand the end of crime, violence, and humiliation for the Portuguese residents of the Sintra area!

> *The Privacy of Citizens . . .*
> Defending citizens' right to privacy . . . Of criminal citizens to be sure . . . But everyone must follow their own agenda . . .

> [Untitled]
> A Friend of every criminal, therefore an enemy of everyone who's decent and upright.

The comments dealing with privacy, and particularly the tone of scandal with which readers refer to the 'privacy of criminals', suggest that privacy is generally considered a non-issue insofar as it is thought not to exist in the public sphere, in what I believe to be a confusion between spatial categories – public and private space – and conceptual categories – public and private sphere. The commenters speak as if it were a contradiction in terms to think of a private sphere within the public space, or as if the concept of public, common and open space were incompatible with the private sphere, whose reach is thought to be exhausted within the boundaries of restricted, confined and intimate spaces. This conception obviously goes hand in hand with the inverse proposition that the public sphere cannot penetrate private spaces, but that would lead to an altogether different debate, whose discussion, while relevant, would force us to stray too far from our focus on the main issue at hand; that is, public video surveillance.

There is another vital issue that must be mentioned here and which could be formulated as a working hypothesis: considering that the notion of privacy effectively gives rise to a multitude of meanings – legal, political, journalistic etc. – we must allow for yet another interpretation of this concept that seems to emerge from the common citizen's perspective, as expressed in these comments. It is the belief that the protection of privacy should not be equally applied to law-abiding citizens – those who should have nothing to fear – and to 'criminals', who hide behind the application of uniform criteria that ultimately provide an umbrella under which the latter's misdeeds are the only thing that benefits from such protection.

It might be overreaching to claim the absolute existence of a commonsense notion that literally divides citizens into two different groups, the righteous and the wicked, and establishes a different set of rights for each, thus creating separate categories of citizenship – first-rate and second-rate citizens. But this notion may be useful here not only because it reflects a distinction expressly conveyed in people's comments regarding categories of citizens, but also because those categories are generally used as an analogy to characterize geographical space. Thus, just as this conception divides citizens into the righteous and the criminal, so does it divide urban space into respectable and disreputable areas, with an equally distinct set of rules, almost as a personification of physical space.

These examples help to explain, firstly, that privacy is commonly understood to possess an almost negative quality, as the realm of everything that is desired to be kept hidden, losing its *raison d'être* in public space for anyone leading a legitimate existence. Whoever is not willing to relinquish the right to privacy in public space, identified as the sphere of everything that ought to be exposed, is thus necessarily doing so to cover up an illegitimate activity.

Secondly, this creates what may be called a dual notion of privacy, comprising a legitimate and an illegitimate kind (the former being the mark of the upstanding citizen and the latter that of the criminal), whose duality is extended to the characterization of different public spaces and whose distinction is closely tied to the rules that are thought to apply in each of those geographical contexts.

But not all readers' comments focus just on the pros and cons of public video surveillance systems, or on reflections surrounding the notion of privacy as a civil right in democratic societies. Other relevant themes include the need for their strengthened presence as a more effective instrument in the war against crime, a line of thought which seems to agree with the security forces' own perceptions and positions. There can also be found comments from readers living outside of Portugal, in places where public video surveillance has existed for several years, such as the United Kingdom, or some cities in France. This last case is particularly interesting because it somehow contradicts the kind of belief that most readers living in Portugal seem to have in the potential virtues of surveillance cameras. This belief, which was described previously as a kind of blind faith, depends as much upon the element of novelty as on the inexperience of those who profess it. The comments of people whose perspective is informed by familiarity with video surveillance in open areas and knowledge of its real impact as a crime-fighting device, serve to demystify such unfounded beliefs. Let us consider some of these readers' opinions.

> *Video surveillance*
> I live in an excellent neighbourhood in the outskirts of Paris, and quite frankly I don't believe video surveillance is useful at all, except maybe to make citizens feel safer, but even that disappears after a few years, before realizing that it was no more than a feeling, because robberies and assaults don't actually stop, and it is almost impossible to immediately make out the perpetrator in a given situation or place, besides which human resources have been gradually cut down on the pretext that they have been allocated to the video surveillance monitoring system. The only solution would be to maintain actual police officers patrolling during the whole day, to have local police with real powers, and to have private police financed by the residents[,] who would also have to be well trained and have actual power to enforce the law.

The following comment follows the same line of thought, this time coming from a reader living in London, who once again underlines the need for greater police patrolling on the streets, regardless of the amount of existing video surveillance, and points out some weaknesses of surveillance devices regarding facial recognition, particularly in cases where perpetrators conceal their faces with masks or hoods.

> *Waste*
> Here in London there are many cameras, which unfortunately are more expensive than a hoodie but are powerless against it. Without police officers on the street,

places in jail ready to accommodate criminals and strict judges, it is impossible to discourage the attraction of an unlawful way of life. Don't spend money on video surveillance, it has already proved to be a failure.

What about the police?
Why don't the police come out from their stations, may I ask? Why don't they go out on the beat instead of sitting on their backside all day scratching their groin behind a desk? There's nothing like the old baton to set those hoodlums straight. We spend money on cameras so that the police can watch hooded criminals' robberies live on TV. In practice the result will be exactly as we have seen in petrol stations. They continue being hit and no one gets caught.

Besides any considerations for or against video surveillance, for or against privacy as the basis of freedom and security, and apart from the scepticism regarding the utility of cameras, these examples reveal an essential belief in the role of active police forces as a key element to solve the problem of crime. In this way, they coincide with the widespread acceptance in Portugal of a model of proximity policing that has been popular since its implementation at the end of the 1990s.

The following comments clearly illustrate the idea repeatedly brought up throughout this chapter that people are not always reacting specifically to the content of the news piece in question, but mostly relating it to their own set of personal values, ideas and beliefs. Note for instance how, out of the three comments on an article reporting that the 'Government authorizes the installation of public video surveillance in Coimbra',[6] not a single one explicitly focuses on the case at hand, but all of them limit themselves to offering general notions about this type of initiative:

Police State
This is how police states are born.

Shame
Yet another blow on our freedom. The illusion of security.

What use has it been?
With all this, I wonder what was the use of the April 25th revolution . . . [. W]e are giving up our freedom.

Once again the importance must be stressed of an element that has been constantly present throughout this work and which is closely related to Portuguese political history, as described in Chapter 1. The experience of a dictatorial regime that lasted almost four decades continues to leave a residual mark on Portugal despite the thirty years of democracy that have elapsed since then.

Whether in popular culture, political discourse or media coverage, references to the dictatorship, fascism, the political police or Salazar are frequently summoned up when any measure is being discussed that somehow involves police forces or intelligence services (still commonly referred to as the 'secret police'), who still carry the burden of being seen as the direct heirs of the dictatorship's repressive machine. This is mentioned here merely to point out that regardless of the actual performance of authorities, whether considered in light of their capacity to act as democratic organisms or in terms of their ability to activate the machinery of control, terms like 'freedom', 'democracy' or 'the Carnation revolution' belong to a vocabulary that refers to struggles of the past rather than to the current state of the modern-day Portuguese democracy.

Conclusion

This chapter began with an explanation of the aim of this research, which was to contribute towards filling what I think is a gap within scientific literature on surveillance, and more concretely on video surveillance, regarding common citizens' perception of these issues. The analysis has been shifted from a description of political decision-making and from the entities directly involved in determining or implementing measures, to an attempt at understanding people's opinions on these themes, how they present their arguments against or in favour of surveillance, and what questions video surveillance raises for them.

One of the most interesting points is the fact that nowhere in all the comments did I find any direct reference to the National Video Surveillance Programme. When people comment on news articles they concentrate on their immediate concerns without reflecting more broadly on the issue as a national strategic policy. This fact shows that there was a failure on the government's part both to communicate and explain this strategy, and to apply it according to its nature; that is, as a nationwide programme, as a coherent and nationally strategic plan. Another thing that becomes clear is that there is a significant element of ambiguity and emotiveness in public perception: security is thought of as a commodity rather than exclusively as a right. In addition, the opinions contained in comments, particularly in the hundreds of comments showing concern for people's safety or more accurately the lack thereof, almost never seem to be based on personal experience, at least to judge from the absence of references to concrete episodes involving the authors of those comments, such as having been victims of any kind of crime. It is fair to suppose that they would exist if that were the case.

What conclusions can be drawn then? That the subjectivity surrounding the perception of security implies that there are some dimensions of this phenomenon which can neither be measured in terms of crime rates or other quantitative parameters, nor solely explained through peoples' direct experience of crime or their assessment of potential danger. As a concept, fear is very difficult to define; as a phenomenon its causes and effects are hard to determine; and as an indicator, it is hard to measure. It can be triggered by multiple situations and vary according to age, gender, social and economic status. But fear is undeniably one of humans' most basic instincts, easily recognizable by anyone who has felt it, and as such, has a very concrete existence. This will be precisely the subject of the next chapter, which will analyse this concept, equating its various dimensions: the inherent subjectivity of fear, the objectivity of numbers and the collective strategies and solutions for the problem of security in cities.

Notes

1 This could be applied to other fields where control and monitoring of individuals is also essential. Consider the forms of surveillance used for commercial purposes. These place much more emphasis on the economic interests that determine actions and consequences (namely the possibility of obtaining greater profits by means of consumer profiling) than on the motivations that drive individuals to 'cooperate' (for instance, taking up the chance to get better bargains in a given chain of supermarkets by obtaining a customer loyalty card) and not question themselves about the potential uses of the personal data that they are sharing (Lace 2005; Zurawski 2011).

2 TSF is a Portuguese national radio whose programming is mainly focused on news content, featuring many debates and forums on current events etc. See http://www.tsf.pt, 28 January 2010.

3 With the heading 'Other personal rights', Article 26 of the Portuguese Constitution, included in the section dedicated to personal rights, freedoms and guarantees (Section II, Chapter I), lists the following items:

> 1. Everyone is accorded the rights to personal identity, to the development of personality, to civil capacity, to citizenship, to a good name and reputation, to their image, to speak out, to protect the privacy of their personal and family life, and to legal protection against any form of discrimination.
> 2. The law shall lay down effective guarantees against the improper procurement and misuse of information concerning persons and families and its procurement or use contrary to human dignity.
> 3. The law shall guarantee the personal dignity and genetic identity of the human person, particularly in the creation, development and use of technologies and in scientific experimentation.
> 4. Deprivation of citizenship and restrictions on civil capacity may only occur in the cases and under the terms that are provided for by law, and may not be based on political motives. (Constitution of the Portuguese Republic, 7th edition. http://www.parlamento.pt/Legislacao/Documents/Constitution7thRev2010EN.pdf)

4 Capital letters are reproduced here as in the original.

5 It should be recalled that Amadora's video surveillance proposal foresaw the installation of 113 cameras throughout several parishes, thus being the project that included the highest number of cameras as well as the largest geographical extension in Portugal.

6 *Público*, 22 December 2008. Due to editorial decisions, it is no longer possible to access the readers' comments.

Chapter 5
The Quest for Security

> Do you want security? Give up your freedom, or at least a good chunk of it. Do you want confidence? Do not trust anybody outside your community. Do you want mutual understanding? Don't speak to foreigners nor use foreign languages. Do you want this cozy home feeling? Fix alarms on your door and TV cameras on your drive. Do you want safety? Do not let the strangers in and yourself abstain from acting strangely and thinking odd thoughts. Do you want warmth? Do not come near the window, and never open one. The snag is that if you follow this advice and keep the windows sealed, the air inside would soon get stuffy and in the end oppressive.
>
> – Zygmunt Bauman, *Community: Seeking Safety in an Insecure World* (2001: 4)

Currently, security is one of the major global concerns and a source of controversial debate worldwide, crossing political, social, economic and cultural spheres. National and international policies have reflected the significance of this phenomenon and the extent of its ramifications, attributing to this concept as many meanings and purposes as the number of actors and spheres of action it involves. It has become synonymous with a defence of the welfare state and the motor behind the thriving of surveillance and protection technologies. At the same time, it influences the internal policies of states while interfering with the very notion of sovereignty and leading to the reformulation of international relations. The manifold notion of security envelops the state and its citizens in a relationship of interdependence and mutual demand brought about by the new risks and threats implied by the increasing globalization of all human action (Beck 1998, 1992).

The terrorist attacks on the United States of 11 September 2001 unquestionably marked a turning point in terms of a global concern with security, triggering a trend of what could be said to be an obsession with security that has become generalized in Western countries. This event had extraordinary repercussions (in the double sense of remarkable and exceptional) in the political, economic and social arenas. Although concern with internal and external security has been present since the origin of sovereign states (Caplan and Torpey 2001; Cole 2001; Crettiez and Piazza 2006; Laniel and Piazza 2006; Torpey 2000), it has gained a new impulse in the last decade, decisively promoting common securitarian strategies among several states. When the government of the United States declared a war on terror, a new paradigm was in effect being introduced, directly affecting the everyday lives of citizens all over the world. It was a policy whose legitimacy was argued on the premise of giving priority to the common good over individual rights and freedoms, implying, often at the expense of the latter, full disclosure of personal information at a variety of levels (Bennett and Lyon 2008; Bigo and Guild 2005; Ericsson 2008; Feldman 2011; Hart 2004; Kapferer 2004a, 2004b; Kapferer and Bertelsen 2009; Lyon 2009, 2003).

The discussion of this complex issue here essentially revolves around a dichotomy that stands out in and is transversal to every debate touching on its related fields, be they political, legislative or academic, that which opposes security and liberty, as Zygmunt Bauman so accurately explains: 'Promoting security always calls for the sacrifice of freedom, while freedom can only be expanded at the expense of security. But security without freedom equals slavery[,] . . . while freedom without security equals being abandoned and lost' (Bauman 2001: 20). While it seems clear that increasing collective security to a certain extent implies the obliteration of individual rights and freedoms, such as free circulation or the right to self-image, private life etc., it might also conversely be claimed that freedom is only assured through security. According to one's perspective, critical knowledge or political sympathies, there are endless variations on this line of thought.

As already stated, when I started doing research for this book, in 2007, my knowledge of video surveillance was limited mainly to contact with other researchers who had published works in other countries, namely in Anglo-Saxon countries. I was convinced, in part because it was the first study of its kind made in Portugal, and in part because the issue of video surveillance had also just started to enter the public debate, that I was going to find an 'easy' terrain in terms of comparing it to other European models. In other words, I presumed that the Portuguese context was a case of the successful importation of foreign models, an idea of which I was quickly disabused. Even though the installation of public video surveillance had indeed been considered in Portugal, the scope of the problems I encountered could not be restricted to a mere comparative study on the uses and effectiveness of this type of public surveillance system.

My first interview with a police officer from the Lisbon headquarters helped to define the direction taken by this work. I was expecting an enthusiastic and even propagandistic defence of the merits of video surveillance, but what I heard from this deputy chief inspector was scepticism and outright opposition to this initiative. And he was not merely referring to video surveillance. He complained mostly of institutional divergences, of conflicting interests and goals, policies that were being imposed from the top down.

There was almost a suggestion of a kind of boycott of what I had (mistakenly) supposed was originally an initiative of this agency. This first interview lasted two hours, during which it was explained to me, by means of diagrams, hypothetical scenarios and with examples, not only that video surveillance in open areas in Portugal was a trend, an enthusiasm with 'modern' technology, but also that it was useless and unrealistic in terms of the actual needs and priorities of this law-enforcement institution.

I was also surprised by this encounter on another level, which had to do with my own preconceptions regarding Portuguese police forces and their officials. Common stereotypes from the past, which largely influenced the idea I had formed in my mind, portrayed the typical police officer as a rude, uncultured and uneducated person. Before me, however, stood a young, 30-year-old officer with a master's degree in political science and an extensive knowledge of current international practices in this field, who was perfectly able to elaborate on the complexities of security, privacy or civil rights and freedoms. Most importantly, I was facing a privileged interlocutor keen to show me how much this institution had evolved; seeking to convey to me that this force, responsible for keeping public order, was not partial to video surveillance precisely because it considered the measure excessively 'controlling'.

As can be learned from the work of Susana Durão (2010) on the history and evolution of the Public Security Police in Portugal, in the years between the end of the 1980s and the second half of the 1990s, there was a major effort by the state to change the previous perception that associated police forces with excessive use of violence, and generally viewed its officers as unprofessional and undertrained. Throughout this decade there was a great investment in providing Portuguese police forces, namely members of the Public Security Police and the National Republican Guard, with better training. Due to their greater contact with the general population, these forces were also those more strongly represented in the popular imagination with the authoritarianism of the previous regime, and thus a symbol of everything the democratic revolution had fought against.

I did not discard the possibility that the tone of this first official conversation conveyed no more than a personal opinion, and thus should not be confused with the official position of the Portuguese Public Security Police. However, despite the danger of overgeneralization, the experience awoke in me an unexpected working hypothesis: was it possible that the apparently hegemonic model for urban security through video surveillance, as witnessed in other European countries, could not be applied in the Portuguese context? If, as the first months of research came to confirm, public video surveillance was not yet effectively operating, what kind of study was I going to make that apparently did not even have an object? After all, what did the National Video Surveillance Programme – whose introduction the government had announced – really amount to if it had not in fact been put into practice anywhere yet? Furthermore, most of the people I spoke with from the different entities involved – Public Security Police, Data Protection Authority, municipal representatives – openly confessed that there was no such thing as a 'programme'; that whatever it was, it certainly could not be called 'national', and that in any case 'video surveillance' in open areas was not even going ahead.

Notwithstanding the coherence and unanimity of these opinions, other relevant questions remained to be answered. Albeit having been informed about the intention to install public video surveillance systems in Lisbon, the first police officer whom I talked to knew little about the situation in other points of the country, although similar procedures were taking place in many different Portuguese cities. This lack of knowledge anticipated what very soon would become obvious: that there was no sharing of information, either internally or among the different state organizations.

One other event confirmed this idea of an existing dissociation of interests. When I first heard of the intention to install video surveillance in Lisbon's downtown – the case of the Baixa Pombalina, described in Chapter 2 – I immediately contacted the local mayor's office to ask for an interview on the subject. To my mind this was a straightforward matter: a project was being drafted by the parish president of that area; it counted on direct support from the Ministry of Internal Affairs, and the Public Security Police's cooperation. It seemed logical that the mayor or someone from his office would have an official position on the subject, since this locale was obviously one of the city's hallmarks. The reply I received on 6 April 2009, in the form of a written statement, was very curious, and not a little baffling. It read as follows:

> On behalf of his Excellency the mayor of Lisbon, I hereby acknowledge receipt of your letter and inform [you] that everything concerning the National Video Surveillance Programme is entirely the government's responsibility. We don't know more about it – in fact we know less – than the MAI [Ministry of Internal Affairs].
>
> His Excellency the mayor has further instructed me to suggest that you should contact the MAI, and that even then should you still need to speak with anyone responsible from this municipality, we can put you in touch with the Municipal Police.

This letter was revealing of something discussed in previous chapters, which becomes very palpable here. As surprising as it may seem, the message I received from the mayor's office showed that the mayor was unaware of the National Video Surveillance Programme's activities, including any proposals currently being made within his own area of governance. Even the tone in which he referred to the Ministry of Internal Affairs was symptomatic of the discordance in institutional relations, and the divergence in political agendas and goals. In other words, the evasiveness of this reply was one of the many instances in which I was confronted with the fact that the national programme was not unanimous, reflecting some idiosyncratic traits of this democratic system, both in terms of its inadequate management of institutional independence, and its inability to gather forces around a unified and global political vision.

After realizing that I would first have to unravel this mesh of intra- and inter-institutional circuits, I knew that my study could not possibly focus strictly on video surveillance as an established phenomenon; that is, on the questions of its efficiency and effects on public life such as they were discussed in so many other works in this field. In fact, these problems were at such an early stage in Portugal, with new developments occurring almost daily, that such data simply did not exist yet. On the other hand, this context provided a unique opportunity to observe the phenomenon from its very inception, witnessing its progression, discussing the different positions and arguments, analysing the successes and failures of a government policy during its implementation period.

At this point it should be clear that video surveillance is at the same time central and peripheral to this study. It is vital insofar as it provides the theme that originates debate and establishes the terms of discussion. It becomes auxiliary because, despite its central role, it is quickly engulfed by other issues that seemed to be more important to my interlocutors, namely their own specific needs and agendas. Video surveillance ended up becoming a mere pretext for broader discussion around the notion of 'security' and the policies adopted in this area. The police were concerned with their logistical problems; citizens were worried about their feeling of insecurity; the Data Protection Authority was interested in defending civil rights; and local politicians were focused on the specific needs of their cities. And in this general discussion, I once again discovered that each party seemed to have a different conception of what was meant by security, and how it should be attained.

Precisely because of the different interpretations of this concept, I wish to address it in its broadest possible sense, freeing it from the strict association with crime and criminality. In this regard, the Portuguese case allows a combination of several perspectives, given that crime rates over the last decade have been shown to be very low and, according to the Annual Report of Internal Security (*Relatório Annual de Segurança Interna*), mostly comprised petty and non-violent criminality. Thus, as will become clear, the question of security crosses different spheres, namely economic and social, and raises issues that range from health and education to urbanism. These will be some of the topics addressed in the following pages.

Barometers of (In)Security

One of the ideas stressed throughout previous chapters is that security strategies, particularly in urban spaces, seek to respond to various concerns: citizens demand to feel safer in the areas where they live, work and enjoy their free time; shopkeepers want their businesses protected; local governments seek to rehabilitate urban areas to make them financially more attractive; and let us not forget the urge of political forces to announce and implement measures which may somehow capture the popular vote. Each of these elements obeys particular interests and follows its own dynamic, making security a multifaceted notion that depends on the combination of a number of factors, coming together under particular circumstances to influence citizens' perceptions of the issue, and at the same time determining the kinds of response given by the state.

In the Portuguese case, it has already been shown that video surveillance does not emerge as a response to any significant shifts in crime numbers or practices, but is included in a strategy of modernization strongly based on technology. Therefore, it does not result from the need to solve an alarming situation in terms of public order and security, but as a strictly political strategy with broader contours. It must also be pointed out that even though the data contained in Table 5.1 does not reveal it, citizens' concern with security is not absent from everyday discourse among the Portuguese population. When asked about their security, people claim to feel unsafe but do not appear to be very confident in the effectiveness of security measures, such video surveillance, as a solution. On the other hand, they do demand intervention in other spheres, namely the improvement of socio-economic conditions.

It is important to try to understand how citizens define 'insecurity', and which measures they consider more efficient to fight it. To this end, the data provided in the *Barómetro da Segurança, Protecção de Dados e Privacidade em Portugal* (Barometer of Security, Data Protection and Privacy for Portugal) could be considered (Table 5.1). This is an annual report on perceptions of security in Portugal, carried out by an independent consulting agency financed by ADT Fire & Security, a company that manufactures and installs security devices, especially video surveillance systems. The survey for the year 2009 (which will be used here as a reference) managed to collect approximately eight hundred valid responses in the areas of Lisbon, Oporto and two other cities in Portugal's northern and southern regions, Viseu and Faro.

What do these data reveal regarding citizens' perceptions of insecurity and their assessment of video surveillance systems? There is one crucial point that stands out from all similar reports and surveys in Portugal: questioned about the country's security, citizens consistently have a pessimistic outlook. For the overwhelming majority, the situation is understood as bad and tending to get worse. However, when asked which measures they deem necessary to invert the situation, video surveillance appears at the bottom of the list.

Surprisingly, on the other hand, people readily recognize the advantages of these systems as a complement to the action of police forces against crime. Besides the fact that this change only occurs when video surveillance is singled out in the question for a specific situation, even then it does not imply that according to the respondents, security problems will be solved by installing surveillance systems rather that their use may aid policing and their presence

Table 5.1: *Data Extracted from the* Barometer of Security, Data Protection and Privacy for Portugal, 2009

How do you describe the evolution of security?	In Portugal: Worse	55%
	In Europe: Worse	60%
	Worldwide: Worse	58%
What are your predictions on the evolution of security for the next 12 months?	Decline	54%
	Improvement	27.5%
How do you rate Portugal in terms of security?	Unsafe	49%
	Safe	41%
How safe do you feel in the city where you live?	Safe	47%
	Unsafe	43%
In terms of safety, how do you feel in your own home?	Safe	68%
	Very safe	25.5%
How do you feel inside your home by comparison to other places?	Safer	64%
	Fairly safe	22%
Of the places you frequent daily, where do you feel least safe?	Public areas (streets)	47%
	Public transport and points of access	14%
	Car parks	12%
Considering Portugal, choose the 3 factors that in your opinion most contribute to the feeling of a lack of safety.[1]	Unemployment	59%
	Changes in the social fabric (immigration)	45%
	New forms of crime (terrorism, organized crime etc.)	39%
How strong do you think is the influence of anti-social behaviour on the feelings of lack of safety?	Strong	54%
	Very strong	26%
Have you ever been the victim of a crime or unlawful act?	No	62%
	Yes	36%
If you have, did you report it to the proper authorities?	Yes	66%
	No	30%
In your opinion, what could be done to promote an atmosphere of greater security in Portugal? (3 measures)[2]	Increase the number of active police	49%
	Introduce/enforce more severe laws	46.5%
	Improve socio-economic conditions	46%
Do you believe that video surveillance can be an aid to policing?	Agree	64%
	Totally agree	19%
Does the presence of a video surveillance system (CCTV) make you feel safer?	Yes	52%
	Indifferent	31%
Do video surveillance devices (CCTV) have a dissuasive effect on illicit behaviour?	Agree	50%
	Disagree	25.5%
Do you consider that video surveillance devices constitute an invasion of privacy?	No	75%
	Yes	20%

Table 5.1: (*continued*)

Would you feel uncomfortable having video surveillance in the following places?	Car par: No	91%
	Public areas (streets): No	87%
	Bank branches and cash machines: No	90%
Indicate the 3 places where you consider it more important to have video surveillance systems in order to increase security or dissuade criminal activity.	Car parks	49%
	Public areas (streets)	47%
	Bank branches and cash machines	44%
Would you give up your privacy in order to install video surveillance in the streets if that meant increasing your feeling of safety?	Yes	77%
	No	18%
Do you feel more afraid in the daytime or at night?	At night	81%
	Indifferent	17%
Is your perception of security altered by the fact that the public transport on which you travel is equipped with video surveillance systems?		48.5%
	It is increased	33.5%
	Not affected	
Is your perception of security altered by the fact that the bars and discos where you go are equipped with video surveillance systems?	It is increased	32%
	Don't know/didn't answer	32%
Do you have any misgivings about being filmed by a video surveillance system?	No	82%
	Yes	16%
Which factors would make you feel more comfortable regarding the use of images recorded by video surveillance systems? (Chose the three most important ones)	Assurance that images are not used for other purposes	89%
	Assurance that images cannot be manipulated	89%
	Security of the place where images are stored	69%
Are you aware of the existence in Portugal of any organism responsible for ensuring the protection of personal data?	Yes	45%
	No	55%

'potentially' improve security, namely in large cities and urban areas. This report suggests, for example, that the 'three factors which most contribute to an increasing lack of security' are unemployment, the rise of violence in society, and new forms of criminality (terrorism, organized crime), even though 60 per cent of the respondents answered that they had never been the victim of a crime. Similarly, improving socio-economic conditions, applying stricter laws and strengthening police presence were pointed out as the most relevant measures to ensure security conditions.

Bearing in mind that these reports are funded by a corporation that it is specialized in building and selling security systems, this survey is mostly interested in people's opinion regarding the installation of public video surveillance and finding out whether they have any concerns in terms of privacy issues. The results clearly show that it is possible neither to demonstrate a relationship between perceptions of danger and actual experience of crime, nor to assume that the use of video surveillance or similar measures would make feelings of insecurity disappear, since according to the data, public safety is considered more dependent on increasing employment and improving general social conditions.

In spite of this, when the results of this survey were released, the conclusions announced in the media were that '[a]pproximately eight out of every ten Portuguese citizens would willingly relinquish some privacy rights in favour of public video surveillance as a means to provide a greater feeling of security', referring to the fact that according to the survey, 70 per cent of respondents feel that video surveillance systems strengthen policing while not threatening their sense of privacy.

It is clear from the data that citizens demand more security, but when looking at their priorities, it appears that the strongest responses associate this notion with economic, professional and social instability. A relationship can be established between social instability and criminality, and therefore when people mention unemployment or economic conditions as a principal cause of insecurity, they are indirectly including the fear of crime – as one of the end results of such conditions – in their perception of insecurity. Nevertheless, surveillance cameras are not conceived to address the causes of insecurity identified as most relevant by the majority, and as such do not understandably figure among the solutions either. For most people, these devices have a very limited sphere of action (crime fighting) and their value can only be assessed in terms of their effectiveness within this specific domain. Therefore, if there is something the data presented here suggest, it is that 'security' is a broad term, one which cannot be understood or addressed solely in terms of crime rates and crime-related issues. Overall, it can be said that perceptions of insecurity cannot be strictly identified with the fear of crime.

It is time now to look more closely at the type of criminality that is being considered here, and try to understand its implications in this context. Here there are nuances which go beyond the mere recording of objective data on crime rates and percentages of reported crimes. The Portuguese *Internal Security Annual Report* classifies different types of crime according to specific features. Among these categories, the so-called 'street crimes', included in the crimes against property, are those that present the highest number of occurrences. This category includes such felonies as car theft, pick-pocketing, drug consumption and minor trafficking. Teresa Heitor points out that 'while a significant percentage of felonies in urban areas can be included in the category of "petty street crimes" or "crimes of opportunity", they affect citizens' well-being and promote the emergence of fear' (Heitor 2003: 32).

Indeed, even though criminality in Portugal is low in terms of reported crimes, its enlargement by the media and through word of mouth makes it gain proportions which security forces themselves do not acknowledge. The work of Teresa Caldeira on São Paulo, Brazil, presents an exhaustive study of inhabitants' narratives and everyday conversation, focusing on 'the talk of crime':

> everyday conversations, commentaries, discussions, narratives, and jokes that have crime and fear as their subject . . . are contagious. Once one case is described, many others are likely to follow. . . . The repetition of histories, however, only serves to reinforce people's feelings of danger, insecurity, and turmoil. Thus the talk of crime feeds a circle in which fear is both dealt with and reproduced, and violence is both counteracted and magnified. It is in such everyday exchanges that opinions are formed and perceptions shaped: that is, the talk of crime is not only expressive but productive. (Caldeira 2000: 19)

In a study carried out in 2010 on perceptions of security in Portimão (a major city in Portugal's southern region popular amongst tourists), Pedro Moura Ferreira and Susana Durão state that the motives which lead to citizen's feeling of insecurity belong to the domain of sensations and appearances. In their words, 'the causes for daytime and night-time fears do not differ greatly, and can be essentially defined as the dread of strangers, of anyone who is regarded, whether for economic, social or ethnic reasons, as being somehow "unusual", and for that reason considered threatening' (Ferreira and Durão 2010: 11). Feelings of insecurity are therefore not dependent exclusively on the types of crime, but also on what may be called a sociographic characterization of fear. On this topic, Bauman remarks:

> It has been, for instance, widely noted that the opinion that the 'world out there' is dangerous and better to be avoided is more common among people who seldom, if ever, go out in the evenings, when the dangers seem to them most terrifying; and there is no way of knowing whether such people avoid leaving their homes because of their sense of danger, or whether they are afraid of the unspoken dangers lurking in dark streets . . . or because, lacking direct personal experiences of threat, they are prone to let their imaginations, already afflicted by fear, run loose. (Bauman 2006: 3)

Even considering elements that are thought to be strictly subjective, such as feeling of fear and insecurity, we cannot neglect the relevance or truthfulness that they have for those who express them. At the same time, we cannot overlook the fact that this same discourse is enmeshed in political and economic strategies (Delanty 2008; Eriksen, Bal and Salemink 2010; Low and Smith 2005). As Goode and Ben-Yehuda note on the idea of moral panic, '[i]n the moral panic, the public fears may be mistaken or exaggerated, but they are real, they do not have to be "engineered" or "orchestrated" by powerful agencies, institutions, bodies, or classes such as the media, the legislature, or the power structure' (Goode and Ben-Yehuda 1994: 129). The so-called global threats also belong to the group of elements which generate fear and insecurity, as Katja Aas explains when she states that terrorism, pandemic disease (swine flu being the most recent example) and organized crime have become 'potent symbols of the spirits of our times' (Aas 2007: 14).

Looking at the Portuguese case, and returning to the *Barometer of Security*, terrorist threats appear as one of the three factors which 'most contribute to the feeling of insecurity'. Note that these forms of threat, and this particularly refers to the terrorist activity that has afflicted countries such as the United States of America, England or Spain, had found absolutely no expression in Portugal. This example helps to illustrate that the relationship between the fear of crime – whatever its nature or dimension – and feelings of insecurity, obeys a logic which cannot solely be explained through objective criteria. Moreover, these phenomena assume specific meanings that shift according to the historical, economic and social contexts in question. In the particular period under analysis here, political action regarding matters of security was determined equally by internal needs and by international resolutions.

National policies must respond to the global situation while at the same time attending to local requirements based on such data as the *Barometer of Security* or surveys on crime victimization. In addition, government action constantly faces party disputes and political opposition which ultimately determines the policy-making involved. As David Garland (2002) observes, much of this is intended to appease opposition and show that the phenomenon of crime is being closely followed and properly dealt with. Forced to install surveillance cameras in open streets, municipalities take the opportunity afforded by this investment to rehabilitate urban spaces, thus increasing the land and property market values of areas that are later publicised as profiting from increased security. When I questioned a member of Parliament about political pressure to respond swiftly to criminal phenomena, which are easily blown out of proportion due to their enormous media impact, she remarked:

> This is a very negative generalization of our society. I understand that there may be other societies with much more violence, but in Portugal I think things should be better adjusted to our particular circumstances. Recently it seems as if every council is applying for video surveillance. I'm not saying it is just a fad, but I think it is being used as the easy answer for extremely complex problems.
>
> Many political leaders are convinced that they have found the solution, but I think they are mistaken. I don't believe that packing our cities with surveillance cameras will make crime go away: it will simply adapt. There should probably be a greater investment in other areas.

As has been shown, investment in technological urban monitoring systems aims to provide security for populations who claim to be frightened – of going out at night and getting mugged – either because they have witnessed such events themselves; because they have heard other people's stories; or simply because they are warned by the media that crime is increasing, and that we live in insecure times (Ferreira and Durão 2010; Frois 2011a). This perception stems both from common claims such as 'when we go out in the street at night there isn't a single soul around' and from the arguments presented by police forces, whose assessment circumscribes periods of most frequent occurrence (the night-time) and identifies underlying causes – depopulation, an ageing population, urban degradation etc.

Thus, the need to see in order to act has become central in political and police arguments in favour of video surveillance. However, as the member of Parliament cited above points out, this rationale tends to concentrate a complex problem in a single solution, often failing to understand that its efficiency depends above all on a broader and more urgent strategy: the need for more efficient urban management and planning. It has already been established both that fear and security have a subjective factor, and that the impact of these phenomena on populations fluctuates according to elements such as age group, personal income and other social variables. It seems clear, however, that in all situations, in terms of both causes (desertification; urban decay) and effects (crime; fear), the question of visibility is of the utmost importance and it cannot be reduced to a matter of monitoring. Community cohesion, accomplished by means of a 'natural surveillance' (Heitor 2003)[3] that is typical of neighbourhood bonds, is part of the urban planning strategy in some areas. As a councillor in charge of civil protection explained to me regarding Coimbra's downtown:

> We [the local council members] had never considered video surveillance as a strategy before because according to the data that reached us from police sources, ours was not a problematic area. But the fact is that its prominence and high density of pedestrian traffic makes it more prone to petty theft. Then there was another problem connected to urban rehabilitation: most buildings in the district are occupied by commerce, so there are very few residents. This creates an ideal setting for shop burglary.
>
> I don't think video surveillance was ever regarded as a solution for this feeling of insecurity, but in order to bring some comfort to the people here, the city council decided to give this measure more serious thought. The solution is to populate these depressed areas. Video surveillance is just a transitional step, a means to an end. When this part of town is fully rehabilitated, the system will no longer be necessary.

Considering that thefts and acts of vandalism were reported daily in this part of town, the local council decided to take measures that would help to diminish criminality, making this distinguished district more attractive to the passers-by who came there to shop, and more importantly, to potential future residents. But this entity always maintained that the measure should be used only during a transition period. As soon as resident concentration increased, police monitoring through cameras should be replaced with residents themselves, whose presence would naturally would take over this function.

Even if the strategy of rehabilitating urban spaces implies resorting to technological devices that strengthen security, these should be not limited to surveillance cameras, and other measures, perhaps more basic and urgent ones, should be considered as well. Among them could be mentioned better street lighting, which has been proven by many studies to be very effective in reducing both crime rates and feelings of insecurity (Pease 1999), or urban planning, which promotes a greater equilibrium between commercial and residential areas, therefore offsetting differences between daytime and night-time crime rates. Whatever feeling of security surveillance cameras may bring, which could be in most cases purely psychological, they do not have the power in themselves to attract populations to a given space and as such should not constitute a primary instrument of urban rehabilitation. The importance of this balance between commercial and residential areas is also supported by current Baixa Pombalina residents, one of whom explained:

> I wouldn't say that I feel afraid in the Baixa Pombalina, but the truth is that I never get out of the car without looking around first; to see if there are many people around or not; that kind of thing. If they close this area to traffic it will become even more deserted and dangerous. I would much prefer it to have better street lighting or the presence of a policeman, even one of those who go unarmed.
>
> I am not at all convinced that a camera can do anything while a person is being assaulted. They should forbid businesses from having their storehouses here. They occupy almost all the buildings' ground floors, so no one lives there. The buildings are in decay while the prices of apartments are overpriced – this area is not prepared for 'family housing'; people use these streets as though they were public toilets, rubbish bins [are] turned over, etc. Do you really think surveillance cameras can do anything about that?

Whether the fear of crime or the feeling of insecurity corresponds to a real or imaginary danger, or more specifically to what crime rates show, the common citizen wishes (and I would say, demands) above all to feel protected. What transpires from the descriptions and arguments given by most of the online commenters cited in the previous chapter is that people are primarily focused on the potential benefits that can result from these technologies, whether it is more security and safety or greater commercial advantages. Debating issues such as personal data protection, the invasion of citizens' private lives and the threat of excessive control or the abuse of power – to name some of the Data Protection Authority's concerns – is considered pointless, misleading and a sign of indifference to the population's welfare.

The discussion should therefore focus on trying to understand whether fear and insecurity derive from media and political hysteria or from a real increase in criminality, giving some thought, for instance, to the fact that in most cases security-force officials are not the ones who ask for this technological device and are more concerned with obtaining other necessary equipment or in reclaiming a more efficient justice system. Thus, it can be assumed that the introduction of video surveillance systems in public areas advances because it is a political response designed to provide a rapid solution to a subjective problem.

It can also be surmised that the perception that such results are hypothetical explains the lack of consistency and coordination in the implementation of these systems. Installing surveillance cameras at a national scale in various regions of Portugal is considered an important measure that will help to rehabilitate cities and strengthen police forces. Yet the absence of prior studies to sustain this claim is blatant: the data presented by so-called security 'barometers' show that the common citizen does not particularly demand video surveillance, as populist discourses conveyed in the media might lead us to believe. Rather, the emphasis is on the need for more police, more employment or less precarious living conditions, for instance (Castel 2007, 2003). All that people want, I was told, is *to feel safer*, which is quite a different thing.

It is precisely this paradoxical element that makes the Portuguese context relevant for a reflection on the importation of foreign models of security and their application in different countries. In a country statistically considered one of the safest in Europe (presenting reduced crime rates that overwhelmingly involve merely petty crime), the push to transform and reconfigure cities to make them safer can be currently attributed to a political reaction to a predominantly subjective phenomenon – fear – rather than to an objective diagnosis of the need of video surveillance devices and means.

Where Danger Comes From

Robert Castel's work *L'insecurité sociale. Qu'est ce qu'être protégé?* (2003) provides an excellent contribution towards framing this debate. This French sociologist thoroughly explores the meanings of security and insecurity and shows that in modern Western societies these notions belong to the sphere of discourse on 'the kinds of protection that a given society adequately guarantees or fails to assure' (Castel 2003: 7). Thinkers such as Zygmunt Bauman have also reflected upon this quest for security, as an ideal which is permanently sought after but never fully accomplished, since it implies finding a balance between freedom and security that does not sacrifice one for the other. That is, of course, when the goal is to create a more just and free society where primacy is given to individual choice and responsibility, but where the state must continue to play a protective role in providing for elements such as health, education, justice etc.; a society that balances the 'state of law' with the 'social state'.

> We are thus in a position to understand why it is the protective economy itself that frustrates the conditions of security whose existence consubstantiates the societies built on the search for that same security. Firstly, [it is] because protective programmes can never be fully accomplished and ultimately produce disappointment and even resentment; secondly[, it is] because its success, albeit partial, in controlling some risks, gives rise to the emergence of new ones. . . . Thus, a reflection on civil and social protection should also lead us to ask ourselves why the current trend of risk aversion makes citizens today never feel fully secure. (Castel 2003: 8)

As Castel explains, this 'risk aversion' is bound up with what may be called a double civic exigency: citizens demand that the state of law fulfil its duty of protection towards individuals and property, while on the other hand that the welfare state guarantee protection systems which will ensure the conditions for individuals to go about their normal activities; that is, working, health, education and other conditions. As for authors like Beck and Giddens, for Castel the notion of risk crucially affects the balance of the equation of security versus insecurity, and he strives to demonstrate how this element unfolds in multiple ways to produce the instabilities in this equation that are found in contemporary societies. Particularly helpful here are this sociologist's observations regarding what he calls the return of uncertainty and the return of the dangerous classes (Castel 2003).

A keen observer of French society, Castel has analysed the case of the 2002 elections, and particularly the expressive results obtained by the extreme-Right party, Front national (National Front). According to Castel, these results reflected a popular reaction to a feeling of having been abandoned by the state in a world that is going through a moment of great instability, where people are constantly alarmed by economic and social shifts and where competitiveness and flexibility are major buzzwords.[4] Castel explains: '[i]f it is possible today to speak of a recrudescence of insecurity, it is mostly due to the existence of sectors in the population who are currently becoming convinced that they are being left behind, helpless to control their future in an ever more changing world' (Castel 2003: 51).

Furthermore, in his point of view migrations become a source of insecurity due either to suspicion towards what is different – Zygmunt Bauman (2001) distinguishes between mixophilia and mixophobia, the former meaning the promotion of cultural blending and the latter the fear of it – or to problems of integration/acceptance between host countries and settling communities. Economic instability and asymmetries emphasize the state's weaknesses; the need for competitiveness makes the work market volatile and precarious (Eriksen, Bal and Salemink 2010); new forms of terrorism are borderless but even so, strongly connoted with particular ethnic and religious communities so that the diffuseness of the fear they generate is quickly channelled through anger towards concentrated focus points.

Against a backdrop of instability marked by social precariousness and uncertainty, fear gains a central position, in a spiral that is fed by diffuse and unknown dangers, often transferred to everything that is perceived as unfamiliar, or even foreign. In other words, there is a tendency in such cycles to resent the powers that be and to look for scapegoats. For instance, the state may be responsible for not being able to guarantee social protection for the unemployed and at the same time held accountable for their existence; unemployment may be seen as resulting from an excessively competitive and specialized work market, and on the other hand this fact blamed on the importation of foreign workers who upset previous working conditions.

But more to the point here, in another work, *La discrimination négative. Citoyens ou Indigenes?* (2007), Castel addresses the problem of the *banlieues*, French suburbs which often become the focus of social problems. Castel claims that the so-called problematic neighbourhoods, areas which in most cities are clearly defined and geographically circumscribed, concentrate all the major elements of insecurity – bad housing, high unemployment rates and criminal activity – and overall can be described as chaotic places which are simultaneously products and producers of social exclusion. Similarly, in the work *Urban Outcasts* (2009), Loïc Wacquant comments on how these suburbs have recently become notoriously connected to the debate in French society on insecurity, and then goes on to make a comparison between Afro-American ghettoes in Chicago and their French counterparts, the working-class banlieues.[5]

Despite these common features, Wacquant seeks to show that not all these suburbs follow identical patterns and tendencies, in the sense that they do not mimic or replicate each other, presenting important differences and specificities. In this particular comparison, he notes for instance that 'the declining urban periphery of France and the African-American ghetto constitute two disparate socio-spatial formations, produced by different institutional logics of segregation and aggregation' (Wacquant 2009: 5).[6]

Returning to the Portuguese case, the work of the sociologist Eduardo Ferreira (1998) can be cited, who has for some time studied crime rates in Portugal throughout the 1980s, 1990s and 2000s. Comparing data obtained from surveys on crime victimization with criminal statistics from the period between 1993 and 2000 allowed him to conclude that 'the growing concern with criminality and the feelings of insecurity attached to it . . . seems to derive from an insoluble contradiction[:] . . . urban violence, which in principle generates high levels of safety concerns, changing our perceptions and experience of city life, seemed mostly to result in acts of vandalism and incivility uncivil behaviour [rather] than in exceptionally violent criminal activity' (Ferreira 2003: 41). Nevertheless, this fear of problematic neighbourhoods also became a prominent issue in the Portuguese context. Indeed, as described in the Introduction, the first political campaign in Portugal to include public video surveillance in its agenda attached this measure precisely to the problem of the so-called critical areas. Although factors such as danger, fear, uncertainty or security depend on multiple variables, there is a general tendency to characterize and personify the agents responsible for them and to combine this with the identification and circumscription of geographical areas that are considered unsafe.

Although the notions of dangerous groups and their connection to particular areas gain more diffuse contours in Portugal, making these elements harder to isolate and their effect less sharply felt, they are also mentioned in at least two public video surveillance projects. It must be stressed out that in Portugal, at least at the institutional level, references that are openly racist or discriminating against particular groups or types of individuals are strongly censured. The rulings proffered by the Data Protection Authority for these two projects, one of which was rejected and the other approved, are a clear example of this strictness. In the former case, that of the project for the Bairro Alto district in Lisbon, which, as has been discussed, is famous both as a nightlife spot and as a traditional or 'typical' neighbourhood, one of the arguments made by local Public Security Police for the installation of surveillance cameras was connected to the alleged presence of 'youth groups from problematic urban areas of surrounding councils such as Amadora, Sintra, Oeiras, and Loures'.[7] However, according to the Data Protection Authority, 'nothing in the reports [in terms of registered crimes] can decisively lead to a conclusive connection between the areas of origin of people who frequent this neighbourhood and the supposedly negative effect of said provenance'.[8]

This is an example which helps to explain how security strategies based on the combination of fear and a perceived lack of safety produce a multifaceted phenomenon that becomes hard to interpret, in terms of distinguishing between facts and subjectivity and between causes and effects, and it ultimately challenges explanations for individual, community or state strategies of protection (Aas, Gundhus and Lomell 2009; Bauman 2006, 2001; Castel 2003; Feldman 2004; Loader and Walker 2007).

The rhetoric used by the politicians – which is intimately intertwined with the dissemination of media discourse – helps to promote a line of action which, whether in itself substantial or empty, tends unquestionably towards populism, and at least in terms of discourse is very effective in the short term. It is efficient in the sense that just by politicians' addressing a 'problem', it is immediately assumed that on the one hand, that problem has been clearly identified, and on the other hand, action is being taken to solve it, even though in reality, proposed actions often are not put into practice, remaining within the realm of intentions – genuine or not.

Fear, Politics, Economy

Although political discourse built around feelings of insecurity on one hand, and the use of video surveillance systems on the other, varies in terms of its formulation and communication according to the different interlocutors and institutions involved, it invariably has to do more with performative aspects (which, as Michel Foucault [1971] demonstrated, are inherent to the act of discourse), than necessarily with concrete measures. In that sense, the divergences and constant undermining that mark the general idea of a National Video Surveillance Programme are only apparently a contradiction of the underlying discourse, and actually constitute a strategy and modus operandi in themselves. As Gusfield observes on what he terms the 'culture of public problems', 'the emphasis is on the way in which ruling groups create legitimation and functional response to their power and interests not by direct assertion of power but by construction of a cognitive and moral reality, a set of motives and directions in the ruled which are consonant with the needs and interests of ruling groups' (Gusfield 1981: 187).

Up to a point, in Portugal the current discourse surrounding security is merely a response to a feeling of discontentment perceived as being shared by vast sectors of the population for which immediate measures and actions seem to be required. Hence the sudden urgency with which the plan for public video surveillance in Portugal emerged. The combination of circumstances is typical of similar phenomena in large urban areas elsewhere: a generalized feeling of insecurity, pressure from professional associations and residents, strategic designs for urban rehabilitation etc.

The difference here is that the process stumbles on successive obstacles that almost appear to be a sign of ill will; a spirit of disinterest, divergence and uncooperativeness which often results in unnecessary debates and irrelevant issues (as some may consider the interference of video surveillance with the right to privacy, for instance). In any case, I believe that ultimately this blockade reveals above all a lack of strategy and planning. In my opinion, the reason for failure to overcome obstacles and reach consensus in this case was mainly connected to a particular political circumstance, namely that the Socialist Party, which held power during the period that is being considered, was not itself totally convinced of its own agenda in this matter. There is a clear indication of this at the lower level of local power, where even those councils which had originally diagnosed a problem in their locality and made public demands for its solution were later reluctant to cooperate actively in the implementation phase, mostly owing to the financial investment that would be required on their part.

The hesitancy that marked the political process – in which even the observance of existing laws seems to have been neglected – exposes a performative aspect which at least partly explains why most public video surveillance projects proposed throughout the years under analysis were frustrated. The failure to learn from experience and the neglect of objective data and sustained arguments, repeating the same mistakes over and over, are clear indicators of the weak foundations on which the whole programme was based from the start.

The elements of fear, politics and economy can be ascribed to all of the parties involved in the process, that both mutually support and manipulate each other (Aas 2007). The people's expression of fear puts pressure on politicians, who respond by acting on the premise that it is better 'to do anything than be accused of doing nothing', as some told me. The price that politicians must pay for appeasing voices that can become a nuisance to governance is usually to do battle with their political opponents. But these political feuds can get even more convoluted, as I discovered in at least in two cities where the instalment of video surveillance cameras was underway. In both these cases, conflict between local and central government was added to the party rivalry, resulting in truly ironic situations, such as members of these municipalities voting against projects conceived by their own party at a central level, just to contradict their opponents at a local level.

Economic interests are obviously also at stake and play a decisive role in the whole process – and this does not only refer the companies that sell the video surveillance equipment. When shop owners demand that municipalities assume responsibility for ensuring safety and providing security in the areas of their businesses, they are also evading their own obligation to acquire independent security systems, such as alarms, barred windows or even simply proper illumination. As several scholars have already demonstrated, the disaggregation and indistinctiveness which contribute to the alienation of inhabitants of urban spaces also promote higher levels of susceptibility to danger (from the other or the unknown) and the search for alternative forms of protection (Bauman 2006; Sennett 1978).

When news of a particular crime breaks, especially when the report focuses on the details of the specific place and time of the occurrence, as often happens through media channels, it instantaneously becomes a talking point among local residents and frequenters of that place and very soon the event is magnified until it becomes the kind of phenomenon that is held responsible for increasing feelings of fear or perceptions of insecurity. This is one of the aspects which the Data Protection Authority has repeatedly underlined in its reports, describing it as the 'fostering of fear', as was discussed in the previous chapter in the context of the Declaration on the European Day of Data Protection. In terms of authorizing the usage of surveillance cameras, making an accurate assessment of whether this apparently generally perceived lack of safety effectively corresponds to reported crime rates and statistics is of paramount importance. In other words, this decision cannot be based on measuring the number of people expressing feelings of fear, but the number of acts that give rise to them.

On the importance of numbers and quantification as objective and reliable data at the service of 'modern modes of government', *Powers of Freedom* by Nikolas Rose is an absolutely relevant reference. According to the author, 'political numbers' are used as a powerful tool of authority by governmental power to calculate tax incomes, predict election numbers or conduct surveys. Numbers can be represented through graphics, charts or statistics; they may consider individual or collective entities (such as business enterprises). Quantifiable data 'appear to depoliticize whole areas of political judgment. They redraw the boundaries between politics and objectivity by purporting to act as automatic technical mechanisms for making judgments, prioritizing problems and allocating scarce resources' (Rose 1999: 198). Bureaucratic and state apparatus resorts to typifications, to borrow James Scott's word, to level out the population and those upon whom it enforces its laws. That is, it tries to create broad categories which aim at 'collapsing or ignoring distinctions that might otherwise be relevant' (Scott 1998: 71).

Political agents are above all interested in responding to the more immediate or practical needs and demands of the population, and they do so by creating the means which may lead to specific measures. In political terms, video surveillance emerges as a solution to fight the 'causes' of insecurity. Police forces and other security entities, on the other hand, are more concerned with the efficiency of these systems as preventative or dissuasive tools, whose greatest potential lies in their use as evidence to obtain convictions. Let us focus on the combination of these two goals to understand how insecurity acts as a catalyst of public policies, and to weigh up the influence of economic or financial factors. According to one resident of the Intendente district in Lisbon (an area that has been the object of several rehabilitation and revitalization projects over the last couple of decades), the installation of surveillance cameras in this locality would have no more significant impact on his everyday life or on the activities taking place there than all the other measures which had already been tried and failed to 'solve' the problem:

> Do you want to know why I think surveillance cameras will be useless there? Because that kind of crime is committed by the drug addict who is desperate and in that moment feels they have nothing to lose. Even when they see the cameras or a policeman in the vicinity they still go ahead or simply try it somewhere else.
>
> Back when Santana Lopes [a previous Lisbon mayor] closed down Intendente, this square did in fact change for a while, but now what we have is a police van on either side of the street, and I don't know how many officers on patrol, so it has become a very hostile atmosphere; but deep down nothing has changed – the result is that prostitutes and drug addicts have just spread out to other adjacent streets, creating an even more dangerous situation.

In this particular area, urban intervention and police presence had effects on the population living in that specific square, but on the whole were ineffective since this just meant shifting criminality around. This particular resident points out yet another significant aspect regarding the type of crime characteristic of this area: prostitution and drug use and trafficking are carried out by persons leading extreme existences, who often are not fit to judge the consequences that their actions may have for themselves or for others.

Finally, on the question of the weight that economic factors have on determining security measures, some members of Parliament agreed that it would be preferable to have more police officers than cameras, but argued that cameras brought one great advantage: they were significantly cheaper than human or other material resources. In addition, the politicians considered that the cameras' fixed and constant presence alone somehow stood in for the uniformed police officer who traditionally performed this task.

It is clear that the economic issue plays a decisive role in the production of security policies in the context of the current reconfiguration of cities from spaces that privilege democratic interaction to vast commercial areas where 'individuals are recast as consumers rather than citizens; as potential harbingers of profit, rather than bearers of rights' (Norris and Armstrong 1999: 8). Katja Franko Aas adds to this that 'while successful consumers are subtly "seduced" to conformity, flawed consumers, who fail to keep up with the demands of the market society, need to be controlled through the use of more repressive control methods such as CCTV' (Aas 2007: 62).

The chain of events witnessed in these cases seems to describe a kind of virtuous/vicious circle in which the ends and the means seem to be mutually justifiable, ultimately perverting both the premises and the aims of this type of security system. The Data Protection Authority, on the other hand, is responsible for ensuring the protection of personal data, as the guardian of democracy against a securitarian drive that it ultimately considers to come dangerously close to totalitarianism, whose ghost still looms from recent history (Pinto 2010a, 2003; Sousa Santos and Nunes 2004).

Officially, the Ministry of Internal Affairs stands behind the project, publicly supporting and encouraging it, but is ultimately aware of its impotence to enforce it, thus rendering its resolve hollow. The Portuguese case shows an attempt to transpose a security model that exists in cities like London, Amsterdam or Lyon (Hempel and Töpfer 2004; Martinais and Bétin 2004; Norris, McCahill and Wood 2004) to another context without previous diagnosis either regarding the latter's specific features and needs, nor the former's legitimacy or ability to meet them. Even more importantly, this attempt seems to completely disregard the feasibility of such an endeavour in terms of logistical and technical matters (Webster 2009).

The *ad hoc*, improvised and uncoordinated character of projects submitted by municipalities often seems to reflect regional competitiveness rather than a national design. There is no other explanation, for instance, for the total lack of communication between Lisbon and Oporto regarding two projects which share many common features, and where the same mistakes were made – a lack of technical means to effectively set up the systems, due mainly to poor planning (Oporto purchased the cameras but was missing a serviceable network to operate them). Nor is there any explanation for the inconsistency of new projects persisting in demanding large numbers of cameras, ignoring the fact that other projects have been turned down precisely due to their lack of feasibility and others are being asked to readjust their proposals to more realistic camera numbers etc.

What the Oporto case shows, is that mere willingness and enthusiasm, which are easily transformed into political and media discourse, do not immediately translate into action. Projects quickly run into a wall of bureaucratic procedures, stalled by the power play enacted among the different parties, which seem more concerned with assigning blame to each other than actually seeing the projects through (Gupta 2006). Added to this, concerns about the legitimacy of such measures, which interfere with sensitive issues such as invasion of privacy and other constitutionally protected civil rights and freedoms, are soon enough also raised by entities such as the Data Protection Authority, which in this case is particularly relevant since it retains final power of decision.

Such clear patterns of behaviour also seem to indicate that there is something radically wrong with the whole process, forcing anyone interested in this phenomenon to return to a basic question: are these surveillance systems really necessary in the first place? Do current crime rates actually justify such measures, especially considering a context where there are so many more pressing and basic needs to attend to? Political discourse helps to draw attention to certain issues and concerns of the population, raising awareness of the need to take action, but it can also help to manipulate the course of that action, and effectively restrict the debate around those issues.

The urgency surrounding video surveillance in public areas in Portugal is a clear example of such an inversion: the media help to construct a feeling of insecurity, which is in turn enhanced by political discourse and finally boosted by certain groups that have a financial or commercial interest in rehabilitating certain urban neighbourhoods or other areas. But attempts to implement such measures seem to be systematically thwarted by the parties who purportedly support it, and a chorus of accusations ensues – frequently bringing back old disputes that have little to do with the matter at hand – bogging the process down in an endless discussion that ultimately paralyses the whole endeavour, and leaving spectators wondering whose idea the whole process was in the first place. More significantly, the endless discussion obscures the issues that are really important rather than helping to clarify them, ultimately casting doubts over the notions of safety, security and privacy in contemporary Portugal.

Notes

1 The other choices presented included: 'the current justice system' (37%); 'increasing violence in society' (33%); 'anti-social behaviour' (31%); 'direct or indirect experience of dangerous situations (theft, mugging, assault, etc.)' (22%); 'mass media treatment of violence' (18.5%); and 'internal security policies' (16%).

2 The other choices presented included: 'better training of security force officials' (32%); 'higher standards of education in the teaching facilities' (32%); 'better conditions/means at the disposal of security forces' (31%); 'installation of video surveillance systems in public areas' (27%); 'increasing the number of private security guards' (13%); 'installing more electronic security systems' (11%); and 'better training of private security guards and agents' (10.5%).

3 Teresa Heitor refers to 'natural surveillance' in the following terms: 'the perception of an area's vulnerability underlies one's avoidance of poorly lit streets or decayed parts of a city. It is generally considered to diminish in areas where informal control is practised, i.e., where residents know one another, watch over each other, and will go to their aid in a dangerous situation' (Heitor 2003: 34).

4 Regarding these issues see the works of Richard Sennett (1998), an author who has thoroughly dealt with changes in the work market under capitalist systems and their social consequences.

5 Wacquant, considering the work of Stanley Cohen (2002, 1989, 1986, 1971), speaks about the 'moral panic over the banlieues', describing 'the suddenness and disproportion of the public reaction to the phenomenon; the designation of a "folk evil", i.e. a wicked category taken to be responsible for the collective ill in question; and

the rise of hostility against this guilty party' (Wacquant 2009: 138). In Wacquant's opinion, the dissemination of moral panic is largely due to journalistic discourse, which demonizes and stereotypes problematic neighbourhoods.

6 The Brazilian example of the *favelas* is a constant reference for studies on urban criminality and violence, and a case study for the different solutions of urban planning to overcome a situation that combines the elements of poverty, drug trafficking, corruption and police violence. Besides Teresa Caldeira's already-cited work (2000), see the volume edited by Machado da Silva (2008) that is entirely dedicated to these issues.

7 Data Protection Authority ruling 68/2009, www.cnpd.pt.

8 Ibid.

Conclusion

Modernization and Backwardness

In 2007, following the format of the BBC television series *Greatest Britons*, Portuguese state television launched the show *Grandes Portugueses* (Greatest Portuguese), in which the audience was asked to determine 'Who was the greatest Portuguese figure of all time?'

The series immediately raised controversy within some sectors for having excluded from its initial list of suggested names the leading figures of the Estado Novo regime, António Oliveira Salazar and Marcelo Caetano. Those responsible for the show in Portugal had considered that the indelible mark left by these two figures in twentieth century Portuguese history nevertheless did not merit their presence within the category of 'Greatest Portuguese'.

The discussion caused by this decision, stirring arguments that exploited the paradox of censorship within a democratic society by appealing to freedom of expression and choice, led to a reversal of the initial intention, eventually including Salazar and Caetano as candidates for the title of 'greatest Portuguese figure of all time'. This change of heart would reveal to be just as polemical, and many political commentators publicly expressed their outrage, considering that it utterly disrespected Portuguese democracy. However, there was one crucial element that had weighed decisively in the resolution: the name of Álvaro Cunhal, the founder and long-standing leader of the Portuguese Communist Party (and probably Estado Novo's most active opponent), also figured in the shortlist. Even though the communist party had made the transition to democracy and been integrated into political and parliamentary life since the revolution of 1974, Álvaro Cunhal and communist regimes on the whole still held an almost equally negative connotation in Portugal as Salazar and fascism did.

The solution thus found, in my opinion, can be described as a kind of Solomonic decision – one in which the role played by the commercial and audience-numbers logic involved in this type of media product were just as important that is, controversy and scandal were not necessarily a hindrance.

Another relevant element that contributed to raising the tone of discussion after the inclusion of Salazar and Caetano was the fact that the series was being broadcast by the state television network, RTP (Rádio Televisão Portuguesa), and given a prime-time slot on its main channel, RTP1.

The debate revolved mostly around the potential consequences that a victory of the former dictator could have. One could argue that this network's consent was in itself evidence of the reluctance to assume a clear position, like the one that had been taken in Germany, for example. Calling on the values of democracy and 'people's' right to choose, using the argument that a confident democracy should not depend on erasing the past, could in this case also be seen as an easy way out of having to take a firm stance, or simply as a purely promotional strategy to secure audience ratings. In any event, the ensuing debates were heated and involved prominent figures of the Portuguese political, academic and cultural elites.

Eduardo Lourenço, the author of one of the most significant works on Portuguese cultural identity, alluded to the ignorance demonstrated by the public on this occasion, and its blatant lack of cultural and historical knowledge. Lourenço and many other intellectuals could not accept that at a time when all of Europe seemed firmly entrenched in democracy, the Portuguese still contemplated the merits of a dictator who had been responsible for the backwardness from which the country had taken so long to recover. According to them, if Salazar were to receive any votes, it would only confirm that this backwardness had not been eradicated after all. However, even those commentators, who saw the possibility of Salazar's victory in the series as an absurdity, pointed out that Portuguese history had always revealed the country's propensity for the figure of the strong leader, of the personality that concentrates absolute power. Moreover, they remarked that this trend did not necessarily reflect so much an admiration of strong leaders as a readiness to relinquish responsibility and leave it up to someone else to look after the running of things and ensure public order. The inclusion of Salazar in this case had to be understood as an integral part of Portuguese cultural identity. In that sense he had undeniably played, whether negatively or positively, a determining role in the history of twentieth century Portugal.

Attempts in these debates to diminish the importance of this episode, by arguing that television programmes had a purely entertainment value and could not be held accountable for political or ideological consequences, were drowned in a discussion that had definitely taken hold of public attention. These revived the age-old Portuguese predicament of a country eternally wedged between backwardness and progress, the centre and the periphery, ignorance and knowledge; of a people torn between profound self-consciousness and self-criticism, and the weakness of a political class that still depended on charismatic and authoritarian figures.

The initial list of one hundred personalities (which included popular singers, film stars, football players and football coaches) was reduced to the final ten candidates, among which were the names of Salazar; Álvaro Cunhal;

Aristides de Sousa Mendes, the Portuguese diplomat who, defying Salazar's orders, had secretly granted thousands of Portuguese visas to Jewish refugees during World War II; Afonso Henriques, the founder and first king of Portugal; Fernando Pessoa, Portugal's most widely known poet; and Luis Vaz de Camões, author of the sixteenth-century epic poem *Os Lusíadas*, which recounts the odyssey of the Portuguese maritime expeditions.

To the bewilderment of many analysts, António Oliveira Salazar emerged as the winner of the show, with 41 per cent of the final votes. This result seemed to confirm some of the theories expounded during the controversy that has been described. Firstly, that Salazar continued to represent political determination, strong vision and authority, by contrast to the absence of values and fickle resolve attributed to the political class that had dominated Portuguese democracy since 25 April 1974; secondly, many have stated that the Portuguese continue to suffer from a serious deficit of democratic and civic awareness.

Regardless of the significance one attributes to this episode, there can be no doubt that it helps to show that the legacy of dictatorship is still present in the collective imagination – from political life to policing, from institutional relations to individual citizens. This analysis of the goals and arguments presented by the different entities involved in the initial stage of the process to implement public video surveillance in Portugal, namely the Data Protection Authority, the Ministry of Internal Affairs and the Public Security Police, concurs with this diagnosis. However, this study has also proved that common perceptions do not establish a direct link between technological devices such as video surveillance systems and the historical legacy of oppression.

This seems to be in contradiction to the previous assertion, but these phenomena function at different levels. Although the activities of many organizations – particularly the police forces – were for some time conditioned by the burden of that history, nowadays its effects operate principally at what has been referred to earlier as a subliminal level, relegated to a residual existence that in fact cannot be said to interfere with the practical affairs of everyday life. This does not mean that the value and resilience of memory in this context are to be underestimated. They are confirmed, for example, by the existence of many forces in Portugal that do not escape a worldview more actively connected with past experience – as in the case of the Data Protection Authority, but also for some members of both left- and right-wing political parties. It simply means that it would be a mistake to confuse those two levels in the analysis of the specific phenomenon in question here. The episode described above of Salazar's election in 2007 as 'the greatest Portuguese of all times' can also help to clarify this point. Despite all the media scandal and debate generated by the event, the fact remains that within a few weeks this epiphenomenon had vanished without a trace, having absolutely no effect on the lives of citizens, much less on political developments or on the democratic system and its institutional functioning.

The campaign of modernization led by Prime Minister Sócrates between 2005 and 2011, following an ideal of progress and innovation whose methods operated precisely on the basis of a radical break with past ways of doing things, was perhaps too quick to dismiss the effects that his 'technological revolution' might have on that level of consciousness in which the political memory continued to interfere. This was particularly true for certain social groups and spheres of power in Portugal whose members had lived under the dictatorship, and for whom historical events continued to guide their activity and modus operandi. For them, the issues of security assume a more abstract and conceptual dimension than the purely immediate concern with safety from the actions of a petty criminal, for instance. On the other hand, and perhaps more significantly, Sócrates' Technological Plan seemed to be unable to cope with the new paradoxes and irresolvable contradictions that are inherent to democratic societies when it comes to matters of security. That is, it failed to acknowledge that while technological devices may be very effective in their strictly instrumental dimension (as tools that help to achieve specific goals and measures), they cannot substitute a reform of mentalities or solve the conundrums generated by contemporary paradigms.

In a country that is strongly marked by the presence of a welfare state system, where liberal or neo-liberal policies are frequently understood as the equivalent of abandoning citizens to their own devices, the debate of security is also divided between these two positions. On the one hand, there is a demand for an active state intervention, whether at the level of the justice system, policing or even the sovereign role of the state in ensuring the protection of citizens – and here the concept is being used in its broadest sense. On the other hand, democracy has brought with it an apparently paradoxical exigency from that same interventionist state: the demand that the freedoms (of expression, action and circulation) and rights (to privacy and image) consecrated by the democratic system and the Portuguese Constitution, are scrupulously respected, while being particularly adverse to coercive and deliberate interventions in the sphere of security.

This contradiction, which is not restricted to the Portuguese context, has not found a solution in the policies of modernization and development – quite the contrary. In my opinion this is due mainly to the fact that the 'modernization' (in this case a technologically mediated one), as it was conceived and implemented during the period under analysis, followed imported models and did not attend to the historical and cultural specificities of the geographical context in which they were applied.

It may be argued that there was no other plausible course of action; that by definition, progress implies breaking with the past, breaking old habits and thought processes, since it seeks precisely to replace them with a new set of values and modus operandi. Perhaps the technological model followed by Sócrates was indeed correct in principle, in that it was the only way to cut loose from a past of backwardness and apathy, with the paternalistic atavisms of old models. Nevertheless, as this book has strived to demonstrate, 'mentalities' (to use a term that is so frequently incorporated in political discourse) cannot be changed by decree. The memory of practices, methods and outlooks endures throughout the generations and inevitably marks present modes of thinking and acting.

Furthermore, ideas of modernization and progress are useful and convenient as political slogans, but they often remain within the sphere of intentions, confined to a political discourse that is essentially rhetorical and is constantly in a process of self-perpetuation and self-destruction. Ultimately, what has been observed here is that the differences between countries and regions, and even the nuances existing within each of them in terms of cultural and social standards, must determine the interpretation of the circumstances that individuals identify themselves as being in. While in some contexts, namely the urban ones studied herein, the question of security assumes great importance, and investment in video surveillance systems is encouraged by local governments, commercial associations and resident populations, in other contexts the concerns with 'security' are more basic, more ontological, so to speak, and individual necessities play a bigger role in a dominion that has more to do with the existing conditions of life.

One of the principal points to retain when reflecting on the issues of privacy, protection of personal data or the activity of the Data Protection Authority pertains to the existence of specific legislation and regulation of these matters. It has been made clear that public and semi-public video surveillance systems are, to a greater or lesser extent, widely present throughout most European capitals. In the United Kingdom, the country where that presence has been particularly intense, the proliferation of cameras was so overwhelming and precipitous that when they emerged in the 1980s, there was almost no legislation to regulate their use. The problems connected with privacy and civil rights only started to be effectively addressed near the end of the 1990s, in what has been described as a race against time. Portugal, on the contrary, has witnessed precisely the inverse of this process.

As has been already explained here, the use of public video surveillance is inscribed within a politically strategic framework of technological innovation that began in 2005, but the laws which regulate the protection of personal data and privacy are much older and prescribe very precise limits in this area. Those parameters were established in the 1976 Constitution for historical and cultural motives, and more precisely because of Portuguese democracy's relative youth, and they have remained unchanged ever since.

After almost fifty years of dictatorship, of the systematic violation of civil rights and freedoms, the new democracy was particularly keen on safeguarding the free exercise of those liberties. The Data Protection Authority, whose consulting functions have a legally binding power in these matters, was in effect bestowed with regulating powers in 1994. The fact that the existence of such an organism preceded the arrival of measures intended to implement public video surveillance in Portugal inverted the conditions that were witnessed in other countries, where the widespread use of technological devices for surveillance came ahead of its specific legislation. The Data Protection Authority in Portugal is responsible for overseeing and ruling on any use that is made of personal data, namely to ensure that the right to privacy, image and free circulation is respected. This point is particularly noteworthy as far as the specificity of the Portuguese context is concerned in comparison to its European counterparts. The legacy of political repression has caused an adoption of positions that are more conservative and cautious in this respect than in countries whose democratic histories have had a different evolution.

The difficulty of making an accurate assessment of the effects of the implementation of video surveillance systems on crime rates within a given geographical area is in all respects similar to – if not actually an inherence of – the difficulty of determining with exactitude the true levels of crime, including those that are not officially reported, those that most contribute towards generating safety concerns, and so on. Any document that seeks to contemplate all these dimensions must necessarily also take into account the economic, social, demographic, political and cultural factors involved. In order to correctly understand whether criminality has increased or decreased in a given area during a specific period – which preferably should be longer than just one year – it is crucial to measure such variables as the changes in population density, shifts in the nature of the main activities developed there (such as commerce, industry, recreation etc.) and transformations in the infrastructures that are present in the area (schools, hospitals, transportation, roads etc.).

All these elements are determinant when putting together a well-grounded examination of the perceptions of security and feelings of insecurity in a given context. As has been discussed, video surveillance systems are designed to perform a role as instruments of crime prevention and deterrence – insofar as they are able to effect a change in behaviours and resultant social relations (and not just in terms of criminal activity) – but they are not meant to act as instruments of rehabilitation or regeneration. The existence of these contradictory propositions has been made explicit throughout this investigation, which has highlighted the doubts, hesitations and interrogations that have marked both the implementation of the National Video Surveillance Programme and the process of installing public video surveillance systems. These doubts were conveyed to me by the vast majority of my interlocutors, namely those who were directly involved in the process; that is, politicians or police officials. As political measures, these systems are intended to somehow appease citizens, business owners and political opposition, demonstrating government concern and its proactive attitude. However, ultimately, they do not tackle, or indeed eliminate, the causes that underlie the instability and precariousness responsible for the conditions which required those measures in the first place.

Afterword

While seeking to move beyond a descriptive account, this work has avoided taking an ideological standpoint or using it to establish a political argument on state control; my approach has avoided taking sides, whether for or against the use of public video surveillance, considering that such perspectives often result in partial or ambiguous analyses. However, I have to agree with Samatas when he says that 'doing surveillance studies is itself an inherently political act directed against authoritarian, anti-democratic and discriminatory practices. In simply "doing" surveillance studies, one is taking a side' (Samatas 2005: 195). There are two aspects that I consider worth noting, and which can help to explain (if only partially), the absence of a definite position on this matter, even considering that the social scientist is not obliged to take an active stance on the social reality that is being studied.

In the first place, during the several years of working on surveillance-related issues in their multiple manifestations in Portugal, I was unable to reach a satisfying answer to the question of whether we can legitimately speak of a deliberate intention behind the use of video surveillance devices to violate the privacy of citizens or any other of their civil rights. This impossibility resulted as much from the empirical data that I was given to observe throughout my fieldwork, as from my interpretation of the existing literature on other contexts. I believe that it is important to distinguish the intentions that motivate the actions leading to a given situation from the consequences that may result from them.

On the other hand, the determinism that marks some of the existing theories on the subject have also revealed themselves to stand on weak foundations, and moreover, to be incompatible with contexts other than their own. The awareness that the problems which underlie the use of surveillance and monitoring technologies must not be treated indiscriminately nor taken out of their context is precisely the reason why this book has abstained from adopting a definite and general position regarding the theme of public video surveillance. Instead, this work has tried to contribute an analysis that includes the largest possible number of interlocutors; that follows the evolution of the whole process, in order to identify and understand the specificity of the Portuguese case, and contribute to more detailed and less generalist interpretations of the phenomenon.

In the second place, one other aspect had a profound influence on this investigation, acting as a binding agent for the questions and goals that this work has sought to respond to – the realization that all the parties involved in this process actually agreed on the need for someone who could observe the process from the outside, so to speak. As my research progressed, I discovered that it was left to me to put together pieces of a puzzle which remained inaccessible to the parties involved – and this inability was actually expressed to me quite openly on several occasions. My privileged position resulted from such simple facts as that I was free to visit the sites where video surveillance systems were being set up, and talk to the different entities and agents responsible for those projects; or that I could gain access to certain documents or rulings, and even to specific legislation – in short, that I was allowed to cross institutional boundaries that those involved in the process were reluctant to overcome, for historical, political or cultural reasons.

Bibliography

Aas, K.F. 2007. *Globalization & Crime*. London: Sage.

Aas, K.F., H.O. Gundhus and H.M. Lomell (eds). 2009. *Technologies of Insecurity: The Surveillance of Everyday Life*. Oxon: Routledge.

Abélès, M. 2010. *The Politics of Survival*. Durham, NC: Duke University Press.

Abrahams, R. 1998. *Vigilant Citizens: Vigilantism and the State*. Cambridge: Polity Press.

Agamben, G. 2005. *State of Exception*. Chicago: University of Chicago Press.

Aoude, I. 2004. 'Arab Americans and Criminalization of Dissent', in B. Kapferer (ed.), *The World Trade Center and Global Crisis.* London and New York: Berghahn, pp. 46–51.

Arriaga, L. 1976. *Breve História de uma Organização Salazarista.* Lisbon: Terra Livre.

Balzacq, T. and S. Carrera. 2006. *Security versus Freedom? A Challenge for Europe's Future*. New York: Routledge.

Bannister, J., N. Fyfe and A. Kearns. 1998. 'Closed Circuit Television and the City', in C. Norris, J. Moran and G. Armstrong (eds), *Surveillance, Closed Circuit Television and Social Control.* Aldershot: Ashgate, pp. 21–39.

Barómetro da Segurança, Protecção de Dados e Privacidade em Portugal. 2009. *ADT Fire & Security*.

Barreto, A. 1994. 'Portugal, a Europa e a democracia', *Análise Social* 29(129): 1051–69.

Barreto, A. 1995. 'Portugal na periferia do centro: mudança social, 1960-1995', *Análise Social* 30(134): 841–55.

Barreto, A. 2002. *Tempo de Incerteza*. Lisbon: Relógio D'Água.

Barreto, A. 2003. 'Social Change in Portugal', in A.C. Pinto (ed.), *Contemporary Portugal: Politics, Society and Culture*. New York: Columbia University Press, pp. 159–82.

Bauman, Z. 2001. *Community: Seeking Safety in an Insecure World*. Cambridge: Polity Press.

Bauman, Z. 2006. *Liquid Fear*. Cambridge: Polity Press.

Beck, U. 1992. *Risk Society: Towards a New Modernity*. London: Sage.

Beck, U. 1998. *World Risk Society*. Cambridge: Cambridge University Press.

Berger, P. and T. Luckmann. 1967. *The Social Construction of Reality: A Treatise in the Sociology of Knowledge*. New York: Anchor Books.

Bermeo, N. 1997. 'Democracy and the Lessons from Dictatorship', *Comparative Politics* 24(3): 273–91.

Bennett, C. and K. Haggerty (eds). 2011. *Security Games: Surveillance and Control and Mega-Events*. London and New York: Routledge.

Bennett, C.J. and D. Lyon (eds). 2008. *Playing the Identity Card: Surveillance, Security and Identification in Global Perspective*. London and New York: Routledge.

Best, J. (ed.). 1989. *Images of Issues: Typifying Contemporary Social Problems*. New York: Aldine de Gruyter.
Bigo, D. 2008. 'Security: A Field Left Fallow', in M. Dillon and A. Neals (eds), *Foucault on Politics, Security and War*. London: Palgrave, pp. 93–114.
Bigo, D. and E. Guild (eds). 2005. *Controlling Frontiers: Free Movement into and within Europe*. Aldershot: Ashgate.
Bigo, D. and A. Tsoukala (eds). 2008. *Terror, Insecurity and Liberty: Illiberal Practices of Liberal Regimes after September 11*. New York: Routledge.
Bird, S.E. 2010. 'The Anthropology of News and Journalism: Why Now?', in S.E. Bird (ed.), *The Anthropology of News and Journalism: Global Perspectives*. Bloomington: Indiana University Press, pp. 1–20.
Boskŏvić, A. (ed.). 2008. *Other People's Anthropologies: Ethnographic Practice on the Margins*. New York and Oxford: Berghahn Books.
Bourdieu, P. 1991. *Language and Symbolic Power*. Cambridge: Polity Press.
Brighenti, A.M. 2010. 'Democracy and its Visibilities', in K. Haggerty and M. Samatas (eds), *Surveillance and Democracy*. London and New York: Routledge-Cavendish, pp. 51–68.
Burgess, P. (ed.). 2010. *Handbook of New Security Studies*. New York: Routledge.
Caldeira, T. 2000. *City of Walls: Crime, Segregation, and Citizenship in São Paulo*. Berkeley: University of California Press.
Calenda, D. and C. Fonio. 2009. 'Gli studi sulla sorveglianza in Italia: alcune riflessioni conclusive', in D. Calenda and C. Fonio (eds), *Sorveglianza e Società*. Rome: Bonanno, pp. 117–23.
Caplan, J. and J. Torpey (eds). 2001. *Documenting Individual Identity*. Princeton: Princeton University Press.
Caplan, P. (ed.). 2003. *The Ethics of Anthropology: Debates and Dilemmas*. London and New York: Routledge.
Castañares, J.C. and A.M. Juste. 2002. 'Spain: In the Centre or on the Periphery of Europe?', in A.C. Pinto and N.S. Teixeira (eds), *Southern Europe and the Making of the European Union. 1945-1980s*. New York: Columbia University Press, pp. 41–63.
Castel, R. 2003. *L'insecurité sociale. Qu'est-ce qu'être protégé?* Paris: Seuil.
Castel, R. 2007. *La discrimination negative. Citoyens ou Indigènes?* Paris: Seuil.
Cohen, S. 1971. *Images of Deviance*. Harmondsworth: Penguin.
Cohen, S. 1989. *Visions of Social Control: Crime, Punishment and Classification*. Cambridge: Polity Press.
Cohen, S. 2002 [1972]. *Folk Devils and Moral Panics: The Creation of the Mods and Rockers*. London and New York: Routledge.
Cohen, S. and A. Scull (eds). 1986. *Social Control and the State: Historical and Comparative Essays*. New York: St. Martin's Press.
Cole, S. 2001. *Suspect Identities: A History of Fingerprinting and Criminal Identification*. Harvard: Harvard University Press.
Cook, T.E. 2005. *Governing with the News: The News Media as a Political Institution*. Chicago: Chicago University Press.
Crettiez, X. and P. Piazza (eds). 2006. *Du papier à la biométrie. Identifier les individus*. Paris: Presses de Sciences, Po.
Cunha, M.I. 2002. *Entre o Bairro e a Prisão: Tráfico e Trajectos*. Lisbon: Fim de Século.

Cunha, M.I. 2008. 'Disciplina, controlo, segurança: no rasto contemporâneo de Foucault', in C. Frois (ed.), *A Sociedade Vigilante. Ensaios sobre Identificação, Vigilância e Privacidade*. Lisbon: Imprensa de Ciências Sociais, pp. 67–80.

Curran, H. and J. Seaton. 2003 [1981]. *Power without responsibility: The Press, Broadcasting, and New Media in Britain.* London and New York: Routledge.

Delanty, G. 2008. 'Fear of Others: Social Exclusion and the European Crisis of Solidarity', *Social Policy & Administration* 42(6): 676–90.

Deleuze, G. 1992. 'Postscript on the Societies of Control', *October* 59: 3–7.

Dillon, M. and A. Neal. 2008. *Foucault on Politics, Security and War.* New York: Palgrave Macmillan.

Douglas, J. 2009. 'Disappearing Citizenship: Surveillance and the State of Exception', *Surveillance & Society* 6(1): 32–42.

Doyle, A., R. Lippert and D. Lyon (eds). 2011. *Eyes Everywhere: The Global Growth of Camera Surveillance*. New York: Routledge.

Durão, S. 2008. *Patrulha e Proximidade. Uma Etnografia da Polícia em Lisboa.* Coimbra: Almedina.

Durão, S. 2010. 'The Social Production of Street Patrol Knowledge: Studying Lisbon's Police Stations', in M. Cools, S. Kimpe, A. Dormaels, M. Easton, E. Enhus, P. Ponsaers, G.V. Walle and A. Verhage (eds), *Police, Policing, Policy and the City in Europe*. The Hague: Eleven International Publishing, pp. 79–112.

Ericsson, R.V. 2008. *Crime in an Insecure World.* Cambridge: Polity Press.

Eriksen, T.H. 1995. *Small Places, Large Issues: An Introduction to Social and Cultural Anthropology*. London: Pluto Press.

Eriksen, T.H., E. Bal and O. Salemink (eds). 2010. *A World of Insecurity: Anthropological Perspectives on Human Security*. London and New York: Pluto Press.

Escobar, A. 2012. *Encountering Development: The Making and Unmaking of the Third World*. Princeton: Princeton University Press.

Feldman, A. 2004. 'Deterritorialized Wars of Public Safety', in B. Kapferer (ed.), *State, Sovereignty, War.* New York and Oxford: Berghahn Books, pp. 16–28.

Feldman, G. 2011. 'Illuminating the Apparatus: Steps toward a Nonlocal Ethnography of Global Governance', in C. Shore, S. Wright and D. Però (eds), *Policy Worlds: Anthropology and the Analysis of Contemporary Power.* New York and Oxford: Berghahn Books, pp. 32–49.

Ferguson, J. 1994. *The Anti-politics Machine: 'Development', and Bureaucratic Power in Lesotho*. Minneapolis: University of Minnesota Press.

Ferguson, J. 2006. *Global Shadows: Africa in the Neoliberal World.* Durham, NC: Duke University Press.

Ferreira, E. 1998. *Crime e Insegurança em Portugal: Padrões e Tendência, 1985–1996.* Oeiras: Celta Editora.

Ferreira, E. 2003. 'Violência e insegurança urbana: um fenómeno em crescimento ou em transformação? O caso da área metropolitana de Lisboa', *Cidades – Comunidades e Territórios* 7: 37–57.

Ferreira, J.M. 2001. 'Os Militares e a Evolução Política Interna e Externa (1974–1982)', in J.M.B. Brito (ed.), *O País em Revolução*. Lisbon: Editorial Notícias, pp. 11–61.

Ferreira, P. and S. Durão. 2010. *A Percepção da Segurança em Portimão*. Relatório de Investigação. Lisboa: Instituto de Ciências Sociais.

Fonio, C. 2007a. 'Surveillance and Identity: Towards a New Anthropology of the Person'. *BSA Conference* 12–14 April 2007, http://www.itstime.it/Approfondimenti/Surveillance%20and%20identity.pdf, accessed 10 June 2012.
Fonio, C. 2007b. *La videosorveglianza. Uno sguardo senza volto.* Milan: Franco Angeli.
Foucault, M. 1971. *L'ordre du discours*. Paris: Gallimard.
Foucault, M. 1995 [1975]. *Discipline & Punish: The Birth of the Prison*. New York: Random House.
Foucault, M. 2003. *Society Must Be Defended: Lectures at the Collège de France, 1975–76*, trans. D. Macey. London: Penguin.
Foucault, M. 2004. *Securité, territoire, population: cours au College de France (1977–1978)*. Paris: Seuil.
Franklin, B. 1999. *Social Policy, the Media and Misrepresentation*. London and New York: Routledge
Freire, A. 2009. 'Ideological Identities in Europe: Comparative Perspective of Portugal, Spain and Greece', in J. Viegas, H. Carreiras and A. Malamud (eds), *Portugal in the European Context, Vol. I. Institutions and Politics*. Lisbon: CIES, Celta, pp. 31–51.
Frois, C. (ed.). 2008. *A Sociedade Vigilante. Ensaios sobre Identificação, Vigilância e Privacidade*. Lisbon: Imprensa de Ciências Sociais.
Frois, C. 2009a. *Dependência, Estigma e Anonimato nas Associações de 12 Passos*. Lisbon: Imprensa de Ciências Sociais.
Frois, C. 2009b. *The Anonymous Society: Identity, Transformation and Anonymity in 12 Step Associations*. Newcastle upon Tyne: Cambridge Scholars Publishing.
Frois, C. 2011a. 'Video Surveillance in Portugal: Political Rhetoric at the Center of a Technological Project', *Social Analysis* 55(3): 35–53.
Frois, C. 2011b. *Vigilância e Poder*. Lisbon: Mundos Sociais.
Frois, C. 2012. 'The Fate of "Backwardness": Portuguese Expectations over Modernisation', *Anthropological Journal of European Cultures* 21(2): 89–113.
Garland, D. 2002. *The Culture of Control: Crime and Social Order in Contemporary Society*. Oxford and New York: Oxford University Press.
Gerholm, T. and U. Hannerz. 1982 'Introduction: The Shaping of National Anthropologies', *Ethnos* 47(1–2): 5–35.
Giddens, A. 1985. *A Contemporary Critique of Historical Materialism. Vol. 2, Nation Station and Violence.* London: Macmillan.
Giddens, A. 1991. *Modernity and Self-identity: Self and Society in the Late Modern Age.* London: Macmillan.
Gil, J. 2004. *Portugal, Hoje. O Medo de Existir*. Lisbon: Relógio d'Água.
Gledhill, J. 2000 [1994]. *Power and its Disguises: Anthropological Perspectives on Politics*. London and New York: Pluto Press.
Goode, E. and N. Ben-Yehuda. 1994. *Moral Panics: The Social Construction of Deviance*. Oxford: Blackwell.
Goold, B. 2004. *Closed-circuit Television and Policing: Public Area Surveillance and Police Practices in Britain*. Oxford: Oxford University Press.
Gupta, A. 1998. *Post-colonial Developments: Culture in the Making of Modern India.* Durham, NC: Duke University Press.
Gupta, A. 2006. 'Blurred Boundaries: The Discourse of Corruption, the Culture of Politics, and the Imagined State', in A. Sharma and A. Gupta (eds), *The Anthropology of the State: A Reader*. Oxford: Blackwell, pp. 211–42.

Gupta, A. and J. Ferguson (eds). 1997a. 'Culture, Power, Place: Ethnography at the End of an Era', in A. Gupta and J. Ferguson (eds), *Culture, Power, Place: Explorations in Critical Anthropology*. Durham, NC: Duke University Press, pp. 1–31.

Gupta, A. and J. Ferguson. 1997b. 'Beyond "Culture": Space, Identity and the Politics of Difference', in A. Gupta and J. Ferguson (eds), *Culture, Power, Place: Explorations in Critical Anthropology*. Durham, NC: Duke University Press, pp. 33–51.

Gusfield, J. 1981. *The Culture of Public Problems: Drinking-Driving and the Symbolic Order*. Chicago: Chicago University Press.

Haggerty, K. and M. Samatas (eds). 2010. *Surveillance and Democracy*. New York: Routledge.

Hart, K. 2004. 'September 11th and after', in B. Kapferer (ed.), *The World Trade Center and Global Crisis*. New York and Oxford: Berghahn Books, pp. 11–17.

Heitor, T. 2003. 'Insegurança em meio urbano: o espaço na mediação de oportunidades delituosas', *Psicologia* 21(2): 31–44.

Hempel, L. and E. Töpfer. 2004. *CCTV in Europe: Final Report*. Berlin: Centre for Technology and Society.

Herzfeld, M. 1992. *The Social Production of Indifference: Exploring the Symbolic Roots of Western Bureaucracy*. Chicago: Chicago University Press.

Holston, J. 1989. *The Modernist City: An Anthropological Critique of Brasília*. Chicago: University of Chicago Press.

Holston, J. 2008. *Insurgent Citizenship: Disjunctions of Democracy and Modernity in Brazil*. Princeton: Princeton University Press.

Inda, J.X. (ed.). 2005. *Anthropologies of Modernity: Foucault, Governmentality and Life Politics*. Oxford: Blackwell.

Jewkes, Y. 2011. *Media & Crime*. London: Sage.

Kapferer, B. (ed.). 2004a. *The World Trade Center and Global Crisis: Critical Perspectives*. New York and Oxford: Berghahn Books.

Kapferer, B. 2004b. *State, Sovereignty, War: Civil Violence in Emerging Global Realities*. New York and Oxford: Berghahn Books.

Kapferer, B. (ed.). 2005. *The Retreat of the Social: The Rise and Rise of Reductionism*. New York and Oxford: Berghahn Books.

Kapferer, B. and B.E. Bertelsen (eds). 2009. *Crisis of the State: War and Social Upheaval*. New York and Oxford: Berghahn Books.

Kerr, I. and V. Steeves. 2008. '"Quem está a tomar conta dos miúdos?" Vigilância online a crianças e pré-adolescentes', in C. Frois (ed.), *A Sociedade Vigilante. Ensaios sobre Identificação, Vigilância e Privacidade*. Lisbon: Imprensa de Ciências Sociais, pp. 229–54.

Khazanov, A. and S. Payne (eds). 2009. *Perpetrators, Accomplices and Victims in Twentieth-Century Politics*. London and New York: Taylor and Francis.

Klauser, F. 2009a. 'Spacial Articulations of Surveillance at the FIFA World Cup 2006™ in Germany', in K.F. Aas, E.O. Gundhus and H.M. Lomell (eds), *Technologies of Insecurity: The Surveillance of Everyday Life*. London: Routledge, pp. 61–80.

Klauser, F. 2009b. '"Lost" Surveillance Studies: A Critical Review of French Work on CCTV', *Surveillance & Society* 6(1): 23–31.

Koskela, H. 2003. '"Cam Era" – the Contemporary Urban Panopticon', *Surveillance & Society* 1(3): 292–313.

Lace, S. (ed.). 2005. *The Glass Consumer: Life in a Surveillance Society*. Bristol: Polity Press.

Laniel, L. and P. Piazza. 2006. 'L'encartement, réponse au terrorisme (France-Grande-Bretagne)?', in X. Crettiez and P. Piazza (eds), *Du papier à la biométrie. Identifier les individus*. Paris: Presses de Sciences, Po, pp. 211–36.

Latour, B. 2010. *The Making of Law: An Ethnography of the Conseil d'État*. Cambridge: Polity Press.

Li, T.M. 2007. *The Will to Improve: Governmentality, Development, and the Practice of Politics*. Durham, NC: Duke University Press.

Lianos, M. 2010. 'Periopticon: Control beyond Freedom and Coercion – and Two Possible Advancements in the Social Sciences', in K. Haggerty and M. Samatas (eds), *Surveillance and Democracy*. London and New York: Routledge-Cavendish, pp. 69–88.

Lima, A. 2003. *Grandes Famílias, Grandes Empresas. Ensaio Antropológico sobre uma Elite de Lisboa*. Lisbon: D. Quixote.

Lima, A. and R. Sarró (eds). 2006. *Terrenos Metropolitanos, Ensaios sobre Produção Etnográfica*. Lisbon: Imprensa de Ciências Sociais.

Loader, I. and N. Walker. 2007. *Civilizing Security*. Cambridge: Cambridge University Press.

Lourenço, E. 2001 [1978]. *O Labirinto da Saudade. Psicanálise Mítica do Destino*. Lisbon: D. Quixote.

Lourenço, E. 2002. *Chaos and Splendor and Other Essays*. Dartmouth: Tagus Press.

Low, S. 2004. 'Urban Fear of Crime and Violence in Gated Communities', in C. Casey and Robert B. Edgerton (eds), *A Companion to Psychological Anthropology*. Oxford: Blackwell, pp. 255–73.

Low, S. and D. Lawrence-Zuñiga (eds). 2003. *The Anthropology of Space and Place: Locating Culture*. Oxford: Blackwell.

Low, S. and N. Smith (eds). 2005. *The Politics of Public Space*. New York: Routledge.

Lucena, M. 1994. 'Notas para uma Teoria dos Regimes Fascistas', *Análise Social* 29(125–126): 9–32.

Lyon, D. 1994. *The Electronic Eye: The Rise of Surveillance Society*. Minneapolis: University of Minnesota Press.

Lyon, D. 2003. *Surveillance after September 11*. Cambridge: Polity Press.

Lyon, D. (ed.). 2006. *Theorizing Surveillance: The Panopticon and beyond*. London: Willan Publishing.

Lyon, D. 2007. *Surveillance Studies. An Overview*. Cambridge: Polity Press.

Lyon, D. 2009. *Identifying Citizens: ID Cards as Surveillance*. Cambridge: Polity Press.

Lyon, D., R. Lippert and A. Doyle. 2011. 'Introduction', in A. Doyle, R. Lippert and D. Lyon (eds), *Eyes Everywhere: The Global Growth of Camera Surveillance*. New York: Routledge, pp. 1–19.

Macdonald, S. 2010. 'Making Ethics', in C. Shore, S. Wright and D. Però (eds), *Ethnographic Practice in the Present*. New York and Oxford: Berghahn Books, pp. 80–94.

Machado da Silva, L. (ed.) 2008. *Vida sob cerco. Violência e rotina nas favelas do Rio de Janeiro*. Rio de Janeiro: Editora Nova Fronteira.

Madeira, J., L. Farinha and I.F. Pimentel. 2007. *Vítimas de Salazar: Estado Novo e Violência Política*. Lisbon: A Esfera dos Livros.

Magalhães, P. 2001. 'Desigualdade, desinteresse e desconfiança: a abstenção nas eleições legislativas de 1999', *Análise Social* 157: 1079–93.

Magalhães, P. 2003. 'A confiança nos parlamentos nacionais: regras institucionais, representação e responsabilização política', *Análise Social* 167: 443–65.

Magalhães, P. 2008. 'Redes sociais e participação eleitoral em Portugal', *Análise Social* 188: 473–504.

Martinais, E. and C. Bétin. 2004. 'Social Aspects of CCTV in France: The Case of the City Centre of Lyons', *Surveillance & Society* 2(2/3): 361–75.

Marx, G. 1988. *Undercover: Police Surveillance in America*. Berkeley: University of California Press.

Marx, G. 2006. 'Soft Surveillance: The Growth of Mandatory Volunteerism in Collecting Personal Information – "Hey Buddy Can You Spare a DNA?"', in T. Monahan (ed.), *Surveillance and Security: Technological Politics and Power in Everyday Life*. New York: Routledge, pp. 37–56.

Mateus, A. 2001. 'O 25 de Abril, a Transformação Política e as Transformações Económicas', in J.M. Brandão Brito (ed.), *O País em Revolução*. Lisbon: Editorial Notícias, pp. 253–75.

Mateus, D.C. 2004. *A PIDE-DGS na Guerra colonial: 1961–1974*. Lisbon: Terramar.

Mattelart, A. 2010. *The Globalization of Surveillance*. Cambridge: Polity Press.

McCahill, M. 1998. 'Beyond Foucault: Towards a Contemporary Theory of Surveillance', in C. Norris, J. Moran and G. Armstrong (eds), *Surveillance, Closed Circuit Television and Social Control*. Aldershot: Ashgate, pp. 41–65.

McCahill, M. and C. Norris. 2003. 'Estimating the Extent, Sophistication and Legality of CCTV in London', in M. Gill (ed.), *CCTV*. Leuester: Perpetuity Press, pp. 51–66.

Melhuus, M., J. Mitchell and H. Wulff. 2010. *Ethnographic Practice in the Present*. New York and Oxford: Berghahn Books.

Meskell, L. and P. Pels (eds). 2005. *Embedding Ethics: Shifting Boundaries of the Anthropological Profession*. Oxford and New York: Berg.

Mitchell, T. 2002. *Rule of Experts: Egypt, Techno-politics and Modernity*. Berkeley and Los Angeles: University of California Press.

Mitchell, J. 2010. 'Introduction', in M. Melhuus, J. Mitchell and H. Wulff (eds), *Ethnographic Practice in the Present*. New York and Oxford: Berghahn Books, pp. 1–15.

Monahan, T. (ed.). 2006. *Surveillance and Security: Technological Politics and Power in Everyday Life*. New York: Routledge.

Morgado, M. 2010. *Autoridade*. Lisbon: Fundação Francisco Manuel dos Santos.

Morlino, L. 2010. 'Authoritarian Legacies, Politics of the Past and the Quality of Democracy in Southern Europe: Open Conclusions', *South European Society and Politics* 15(3): 507–29.

Mosse, D. 2011. 'Politics and Ethics: Ethnographies of Expert Knowledge and Professional Identities', in C. Shore, S. Wright and D. Però (eds), *Policy Worlds: Anthropology and the Analysis of Contemporary Power*. New York and Oxford: Berghahn Books, pp. 50–67.

Nery, I. 2004. *Política e Jornais: Encontros Mediáticos*. Oeiras: Celta.

Norris, C. and G. Armstrong. 1999. *The Maximum Surveillance Society: The Rise of CCTV*. Oxford and New York: Berg.

Norris, C., M. McCahill and D.M. Wood. 2004. 'Editorial. The Growth of CCTV: A Global Perspective on the International Diffusion of Video Surveillance in Publicly Accessible Space', *Surveillance & Society* 2(2–3): 110–35.

Palacios Cerezales, D. 2003. *O Poder Caiu na Rua: Crise de Estado e Acções Colectivas na Revolução Portuguesa*. Lisbon: Imprensa de Ciências Sociais.

Palacios Cerezales, D. 2010. 'Repressive Legacies and the Democratisation of Iberian Police Systems', *South European Society and Politics* 15(3): 429–48.

Palacios Cerezales, D. 2011. *Portugal à Coronhada*. Lisbon: Tinta-da-China.

Pease, K. 1999. 'A Review of Street Lighting Evaluations: Crime Reduction Effects', *Crime Prevention Studies* 10: 47–76.

Pels, P. 2000. 'The Trickster's Dilemma: Ethics and the Technologies of the Anthropological Self', in M. Strathern (ed.), *Audit Cultures: Anthropological Studies in Accountability, Ethics and the Academy*. London and New York: Routledge, pp. 135–72.

Pimentel, I.F. 2000. *História das Organizações Femininas no Estado Novo*. Rio de Mouro: Círculo de Leitores.

Pimentel, I.F. 2007. *A História da Pide*. Lisbon: Círculo de Leitores.

Pina Cabral, J. 1991. *Os Contextos da Antropologia*. Lisbon: Difel.

Pina Cabral, J. 2002. *Between China and Europe: Person, Culture and Emotion in Macao*. London and New York: Berg.

Pina Cabral, J. 2005. 'The Future of Social Anthropology', *Social Anthropology* 13(2): 119–28.

Pina Cabral, J. 2006. 'Reflexões finais', in A.P. Lima and R. Sarró (eds), *Terrenos Metropolitanos. Ensaios sobre Produção Etnográfica*. Lisbon: Imprensa de Ciências Sociais.

Pina Cabral, J. 2008. 'Prefácio', in C. Frois (ed.), *A Sociedade Vigilante. Ensaios sobre Identificação, Vigilância e Privacidade*. Lisbon: Imprensa de Ciências Sociais.

Pinto, A.C. 2003. 'Twentieth Century Portugal: An Introduction', in A.C. Pinto (ed.), *Contemporary Portugal: Politics, Society and Culture*. New York: Columbia University Press, pp. 1–46.

Pinto, A.C. 2010a. 'The Authoritarian Past and South European Democracies: An Introduction', *South European Society and Politics* 15(3): 339–58.

Pinto, A.C. 2010b. 'Coping with the Double Legacy of Authoritarianism and Revolution in Portuguese Democracy', *South European Society and Politics* 15(3): 395–412.

Pinto, A.C. and N.S. Teixeira. 2002. 'From Africa to Europe: Portugal and European Integration', in A.C. Pinto and N. S. Teixeira (eds), *Southern Europe and the Making of the European Union*. New York: Columbia University Press, pp. 3–40.

Ponte, C. 2004. *Leituras das Notícias. Contributos para uma Análise do Discurso Jornalístico*. Lisbon: Livros Horizonte.

Pratten, D. and A. Sen (eds). 2007. *Global Vigilantes*. London: Hurst.

Razac, O. 2008. *Avec Foucault, après Foucault. Disséquer la société de contrôle*. Paris: L'Harmattan.

Relatório Anual de Segurança Interna. 2008. Lisbon: Gabinete do Secretário-Geral do Sistema de Segurança Interna.

Ribeiro, G.L. and A. Escobar (eds). 2006. *World Anthropologies. Disciplinary Transformations Within Systems of Power*. Oxford and New York: Berg.

Rosanvallon, P. 2008. *Counter-democracy: Politics in an Age of Distrust*. Cambridge: Cambridge University Press.

Rosas, F. 1989. 'Salazar e o salazarismo: um caso de longevidade política', in AAVV (ed.), *Salazar e o Salazarismo*. Lisbon: D. Quixote, pp. 23–31.

Rosas, F. 2001. 'O Salazarismo e o Homem Novo: Ensaio sobre o Estado Novo e a Questão do Salazarismo', *Análise Social* 35(157): 1031–54.

Rose, N. 1999. *Powers of Freedom: Reframing Political Thought*. Cambridge: Cambridge University Press.

Rowlands, M. 2004. 'Ways in a Heritage Debate', in B. Kapferer (ed.), *The World Trade Center and Global Crisis*. London and New York: Berghahn, pp. 52–58.

Rule, J. 1973. *Private Lives and Public Surveillance*. London: Allen Lane.

Sætan, A.R., H.M. Lomell and C. Wiecek. 2004. 'Controlling CCTV in Public Spaces: Is Privacy the (Only) Issue? Reflections on Norwegian and Danish Observations', *Surveillance & Society (CCTV Special)* 2(2/3): 396–414.

Salgado Matos, L. 2001. 'A Igreja na Revolução em Portugal (1974–1982)', in J.M.B. Brito (ed.), *O País em Revolução*. Lisbon: Editorial Notícias, pp. 63–131.

Samatas, M. 2004. *Surveillance in Greece: From Anticommunist to Consumer Surveillance*. New York: Pella.

Samatas, M. 2005. 'Studying Surveillance in Greece: Methodological and Other Problems Related to an Authoritarian Surveillance Culture', *Surveillance & Society* 3 (2/3): 181–97.

Samatas, M. 2008. 'From Thought Control to Traffic Control: CCTV Politics of Expansion and Resistance in Post-Olympics Greece', *Sociology of Crime, Law and Deviance* 10: 345–69.

Scott, J. 1990. *Domination and the Arts of Resistance: Hidden Transcripts*. New Haven: Yale University Press.

Scott, J. 1998. *Seeing Like a State: How Certain Schemes to Improve the Human Condition Have Failed*. New Haven: Yale University Press.

Sennett, R. 1978. *The Fall of Public Man*. London: Penguin Books.

Sennett, R. 1992. *The Conscience of the Eye: The Design and Social Life of Cities*. New York: Norton.

Sennett, R. 1998. *The Corrosion of Character: The Personal Consequences of Work in the New Capitalism*. New York: Norton.

Shore, C. 2002. 'Introduction: Towards an Anthropology of Elites', in C. Shore and S. Nugent (eds), *Elite Cultures: Anthropological Perspectives*. New York: Routledge, pp. 1–21.

Shore, C. and S. Nugent (eds). 2002. *Elite Cultures: Anthropological Perspectives*. London and New York: Routledge.

Shore, C. and S. Wright. 1997. 'Policy: A New Field of Anthropology', in C. Shore and S. Wright (eds), *Anthropology of Policy: Critical Perspectives on Governance and Power*. New York: Routledge, pp. 3–39.

Shore, C. and S. Wright. 2011. 'Conceptualising Policy: Technologies of Governance and the Politics of Visibility', in C. Shore, S. Wright and D. Però (eds), *Policy Worlds: Anthropology and Analysis of Contemporary Power*. New York and Oxford: Berghahn Books, pp. 1–31.

Shore, C., S. Wright and D. Però (eds). 2011. *Policy Worlds: Anthropology and Analysis of Contemporary Power*. New York and Oxford: Berghahn Books.

Smith, G. 2004. 'Behind the Screens: Examining Constructions of Deviance and Informal Practices among CCTV Control Room Operators in the UK', *Surveillance & Society* 2(2/3): 376–95.

Sousa Santos, B. 2009. 'Tales of Being and Not Being', *Portuguese Literary & Cultural Studies*: 1–46.

Sousa Santos, B. and J.A. Nunes. 2004. 'Introduction: Democracy, Participation and Grassroots Movements in Contemporary Portugal', *South European Society and Politics* 9(2): 1–15.

Spencer, J. 2007. *Anthropology, Politics and the State: Democracy and Violence in South Asia*. Cambridge: Cambridge University Press.

Steiner, J., A. Bächtiger, M. Spörndli and M.R. Steenbergen. 2004. *Deliberative Politics in Action: Analyzing Parliamentary Discourse*. Cambridge: Cambridge University Press.

Strathern, M. 2000. 'Introduction: New Accountabilities', in M. Strathern (ed.), *Audit Cultures: Anthropological Studies in Accountability, Ethics and the Academy*. London and New York: Routledge, pp. 1–18.

Torpey, J. 2000. *The Invention of the Passport: Surveillance, Citizenship and the State*. Cambridge: Cambridge University Press.

Viegas, J. and S. Faria. 2009. 'Political Participation: The Portuguese Case from a European Comparative Perspective', in J. Viegas, H. Carreiras and A. Malamud (eds), *Portugal in the European Context, vol. I. Institutions and Politics*. Lisbon: CIES, Celta, pp. 53–67.

Wacquant, L. 2009. *Urban Outcasts: A Comparative Sociology of Advanced Marginality*. Cambridge: Polity.

Webster, W. 2004. 'The Diffusion, Regulation and Governance of Closed-Circuit Television in the UK', *Surveillance & Society* 2(2/3): 230–50.

Webster, W. 2009. 'CCTV Policy in the UK: Reconsidering the Evidence Base', *Surveillance & Society* 6(1): 10–22.

Westin, A. 1967. *Privacy and Freedom*. New York: Atheneum.

Wood, D.M. and W. Webster. 2009. 'Living in Surveillance Societies: The Normalisation of Surveillance in Europe and the Treat of Britain's Bad Example', *Journal of Contemporary European Research* 5(2): 259–73.

Wright, S. 2011. 'Studying Policy: Methods, Paradigms, Perspectives', in C. Shore, S. Wright and D. Però (eds), *Policy Worlds: Anthropology and the Analysis of Contemporary Power*. New York and Oxford: Berghahn Books, pp. 27–31.

Zedner, L. 2007. 'Pre-crime and Post-criminology?', *Theoretical Criminology* 11(2): 261–81.

Zedner, L. 2009. *Security*. New York: Routledge.

Zurawski, N. 2007. 'Video Surveillance and Everyday Life', *International Criminal Justice Review* 17(4): 269–88.

Zurawski, N. 2010. '"It is all about perceptions": Closed-Circuit Television and Perceptions of Space – What the People Say', *Security Journal* 23: 259–75.

Zurawski, N. 2011. 'Local Practice and Global Data: Loyalty Cards, Social Practices, and Consumer Surveillance', *Sociological Quarterly* 52(4): 509–27.

Zureik, E., D. Lyon and Y. Abu-Laban. 2011. *Surveillance and Control in Israel/Palestine: Population, Territory, and Power*. London and New York: Routledge.

Zureik, E. and M. Salter (eds). 2005. *Global Surveillance and Policing: Borders, Security, Identity*. Portland, OR: Willan Publishing.

Index